"This book provides insight into the unique role that capital cities can and should play as leaders in sustainability. The authors show persuasively that capital cities are different from other types of cities: they are home to the national government, numerous networks of NGSOs, media and think-tanks, and are linked though embassies to the wider global community. Capital cities can – potentially – stimulate the process of developing sustainability solutions in novel ways which in turn can be disseminated throughout the rest of the country, and the international community."

Lisa Benton-Short, Professor of Geography,
George Washington University, USA

"This book presents a conceptually stimulating and data-rich insight into the sustainability policies and practices in the national capitals throughout the world. Does a well-functioning capital city hub necessarily lead to effective sustainability efforts? Is sustainability agenda in the capitals enabled or constrained by their political role? What are the factors of success or failure? The book makes a first attempt to answer these questions by placing the national capitals on the global map of urban sustainability."

Andrey N. Petrov, President, International
Arctic Social Sciences Association and
Associate Professor of Geography,
University of Northern Iowa, USA

Capital Cities and Urban Sustainability

Capital Cities and Urban Sustainability examines how capital cities use their unique hub resources to develop and disseminate innovative policy solutions to promote sustainability. Cities are taking a leading role in defining a sustainable future at a time when national, state, and regional governments in several countries do not provide sufficient leadership.

Capital cities stand out among cities as likely leading drivers in the effort to empower sustainable innovation as they provide a hub for connecting a variety of key constituencies. While acknowledging the successes capital cities have achieved, the international, multi-disciplinary contributors to this work discuss how there is room to do more and improve. The promotion of specific sustainability policies in crucial areas such as clean water provision, high tech innovation, public procurement contracting, and improving flood control in capital cities is examined through various global case studies. The examples range from relatively rich capital cities, such as Copenhagen, where the well-financed hub would be expected to succeed in generating sustainable policies, to poorer cities such as Phnom Penh, where such an optimistic outcome can seem less likely.

Robert W. Orttung is Research Director at Sustainable GW and Research Professor at the Elliott School for International Affairs, the George Washington University, Washington, DC, USA.

Advances in Urban Sustainability

Urban Sustainability in Theory and Practice
Circles of sustainability
Paul James

Sustainability Citizenship in Cities
Theory and practice
Edited by Ralph Horne, John Fien, Beau B. Beza, Anitra Nelson

Capital Cities and Urban Sustainability
Edited by Robert W. Orttung

Capital Cities and Urban Sustainability

Edited by
Robert W. Orttung

Routledge
Taylor & Francis Group
LONDON AND NEW YORK

earthscan
from Routledge

First published 2019
by Routledge
2 Park Square, Milton Park, Abingdon, Oxon OX14 4RN

and by Routledge
605 Third Avenue, New York, NY 10017

First issued in paperback 2020

Routledge is an imprint of the Taylor & Francis Group, an informa business

British Library Cataloguing-in-Publication Data
A catalogue record for this book is available from the British Library

Library of Congress Cataloging-in-Publication Data
Names: Orttung, Robert W., editor.
Title: Capital cities and urban sustainability / edited by Robert W. Orttung.
Description: 1 Edition. | New York : Routledge, 2019. | Includes bibliographical references and index.
Identifiers: LCCN 2018056347 | ISBN 9781138387836 (hardback : alk. paper) | ISBN 9780429426049 (e-book)
Subjects: LCSH: Cities and towns—Growth—Case studies. | City planning—Environmental aspects—Case studies. | Urban ecology (Sociology)—Case studies. | Capitals (Cities)—Economic aspects—Case studies.
Classification: LCC HT151.C3427 2019 | DDC 307.1/216—dc23
LC record available at https://lccn.loc.gov/2018056347

ISBN 13: 978-0-367-73101-4 (pbk)
ISBN 13: 978-1-138-38783-6 (hbk)

Typeset in Goudy
by Apex CoVantage, LLC

Contents

Figures

Tables

Contributors

Alexandra Andhov, Assistant Professor of Corporate Law, Centre for Enterprise Liability (CEVIA), Faculty of Law, University of Copenhagen, Denmark

Marta Andhov, Assistant Professor in Public Procurement Law, Centre for Enterprise Liability (CEVIA), Faculty of Law, University of Copenhagen, Denmark

Sahar Attia, Professor of Architecture and Urban Design at the faculty of Engineering, Cairo University, Egypt and Board Member, Ecocity Builders

Anthony Gad Bigio, Urban Advisor, Adjunct Professor of Sustainable Urban Planning, College of Professional Studies, the George Washington University, Washington, DC, USA

Nicole Forsberg, Program and Community Manager, KTH Royal Institute of Technology, Stockholm, Sweden

Anna Helm, Director, GW Center for International Business Education & Research (GW-CIBER) and Assistant Teaching Professor of International Business, the George Washington University, Washington, DC, USA

Christina Johannsen, Editorial Assistant at the George Washington University, Washington, DC, USA

Jennifer Lessick, Research Associate, Sustainable GW, the George Washington University, Washington, DC, USA

Nguyễn Ngọc Lý, Founder and Director, Center for Environment and Community Research, Hanoi, Vietnam

Robert W. Orttung, Research Director, Sustainable GW, and Research Professor of International Affairs, Elliott School of International Affairs, the George Washington University, Washington, DC, USA

Stanislav E. Shmelev, Director, Environment Europe, Oxford, UK and Visiting Lecturer, University of St Gallen, Switzerland

Irina A. Shmeleva, Associate Professor, ITMO University, St Petersburg, Russia and Director, Institute of Sustainable Development Strategy

Costis Toregas, Director of the Cyber Security and Privacy Research Institute at the George Washington University, Washington, DC, USA, and Board Member, Ecocity Builders

Harold (Hal) Wolman, Professor Emeritus of Political Science and Research Professor, George Washington Institute of Public Policy, at the George Washington University, Washington, DC, USA

Linda J. Yarr, Research Professor of Practice of International Affairs, the George Washington University, Washington, DC, USA

Part I

Theoretical overview

1 Capital cities as innovators in sustainability policy and practice

Robert W. Orttung

Are national capitals meeting their potential?

The research question at the core of this book is: Are national capital cities reaching their potential as hubs for innovation in pushing forward sustainability policies and practices? Sustainability policies and practices in the context of this book are any efforts to develop cities in ways that meet the needs of current residents without compromising the capacity of future generations to live at a similar or better level. This book starts with the assumption that capital cities should be leaders in sustainability policy because they bring together a unique set of actors, who – through networks developed in the physical place of the capital city – should be able to initiate, implement, and disseminate innovative sustainability policies. This collective study considers the unique resources gathered together in a variety of capital cities around the world and inquires to what extent these cities are deploying the resources concentrated in local metropolitan regions in ways that take sustainability seriously (Portney, 2013).

Capital cities form a relatively exclusive club. At the beginning of 2018, United Nations membership consisted of 193 countries, each with a capital city, though Bolivia, Equatorial Guinea, and Swaziland have two, and South Africa boasts three. Today there are many more countries than existed just 100 years ago. In 1900, there were only about 40 nation-states, and most of these were in Europe. At that time, many of the new countries were in Latin America thanks to the breakup of the Spanish and Portuguese empires. The twentieth century saw rapid growth in the number of countries as the Austro-Hungarian, German, and Ottoman empires collapsed after WWI, the British and French empires disintegrated after WWII, and the Soviet Union and Yugoslavia fell apart at the end of Cold War (Gordon, 2006). With the imperial era receding, few new countries are expected to appear in the foreseeable future, meaning that the current number of capitals is stable.

Capital cities as a class have received much less analytical attention than global cities, smart cities, green cities, and a host of other urban categories. When scholars do focus on them, they often stress the symbolic importance of the capitals in the life of the nation. The U.S. capital of Washington, DC has often been seen as sui generis and therefore relegated to its own niche of scholarship separate from the mainstream of studies about urban life (Klemek, 2016; Mayer, Sager, Kaufmann, & Warland, 2016). To put capital cities into the broader context of urban studies, Hal Wolman's Chapter 2 in this volume explains how capital cities are different from other types of cities and how they vary among themselves.

While we still live in a world of countries, cities are taking an important role in creating solutions for a sustainable future at a time when national, state, and regional governments do not provide sufficient leadership (Barber, 2013; Katz & Bradley, 2013). National capital cities stand out among all cities as potential leaders in the effort to empower sustainability innovation because they provide a hub for connecting a variety of key constituencies.[1] By definition, capital cities host their countries' national governments and the related institutions attract numerous international, national, and local stakeholders who use the physical space of the capital cities to lobby for preferred policies. During this process, they can connect and work together to solve critical sustainability challenges. Beyond the national politicians and civil servants, key stakeholders in the capital cities include the capital city government, national non-governmental organizations (NGOs, including industry trade associations, professional associations, labor unions, and environmental groups), corporations and lobby groups, embassies, international organizations, media, university professors, think tank researchers, and the large number of tourists who visit capital cities each year (Figure 1.1). By assembling so many different constituencies in a small place, capital cities have the ability to stimulate the process of developing sustainability solutions in ways

Figure 1.1 The capital city hub

that no other cities are able to achieve. *Capital Cities and Urban Sustainability* explores what capital cities have achieved and where they are not living up to their potential.

This book advances our knowledge of both urban and sustainability issues and distinguishes itself from existing studies by building on the small, but growing, literature examining capital cities to show how these high-profile urban areas are stimulating sustainability policies within their own countries and internationally. The key innovation of the book is in highlighting the capital city's distinctive hub interactions and the role of this ongoing networking to generate new policies and practices aimed at increasing sustainability.

Capital city networks as generators of sustainability policy and practice

The central hypothesis at the core of this book is that the unique collection of actors and resources in capital city hubs help capital city governments produce and disseminate innovative sustainability policies and practices. The dependent variable of sustainability policy and practices can be measured in two ways: first, achieving objective economic, environmental, and equity; and second, active, measurable efforts to pass sustainability policies and implement sustainability practices, which can be summarized in the idea of taking sustainability seriously (Portney, 2013). The independent variable is the hub created by capital cities that brings together institutions, groups, and networks. The following discussion lays out our two variables in more detail.

Dependent variable

Ultimately, this book is concerned with capital city policies and practices that promote sustainability. The case studies focus on specific capital city policies, whether they are protecting lakes or the tree canopy in Hanoi by encouraging greater citizen involvement, or promoting sustainable outcomes through public procurement policies in Copenhagen.[2] The topics and authors are necessarily multi-disciplinary in order to capture the complexity of the capital cities themselves and the large variety of sustainability policies pursued.

As with intangible concepts like democracy, justice, and innovation, the substance of sustainability can be hard to define and measure. We start from the assumption that there is a concrete core to sustainability and work to quantify its key components. The conceptual roots of sustainability date back to the late seventeenth century and have evolved over time (Caradonna, 2014, p. 6). Contemporary definitions for sustainable cities focus on using resources in a way that does not impinge on future generations (World Commission on Environment and Development, 1987). Sustainability science seeks to view the world in big picture terms and understand how the various components "depend on one another, interact, and co-evolve" (Matson, Clark, & Andersson, 2016, p. 4). We implement these ideas by seeking to integrate concern for the environment with a broader understanding that communities need a thriving economy and jobs as well as measures of social justice in order to thrive. The UN Sustainable Development Goals, adopted in 2015, provide the most recent and succinct statement of these objectives, including Goal 11 for sustainable cities and communities.[3]

Portney argues that the existing indexes designed to measure urban sustainability tend to mix policy measures with outcome indicators (Portney, 2013, p. 41). He calls for creating an Index of Taking Sustainability Seriously, with the caveat that there still is not enough empirical data to state with confidence how much specific actions, policies or programs influence objective measures of sustainability. Nevertheless, cities that "take sustainability seriously" are hopefully making progress toward greater sustainability.

We implement the measurement of the dependent variable – sustainability policies and practices – in the following way. Chapter 3 in this volume by Stanislav E. Shmelev and Irina A. Shmeleva provides a quantitative approach by measuring the performance of selected capital cities in terms of their carbon dioxide emissions. Their analysis grounds the subsequent city case studies with a simple and straightforward metric. The case studies that make up the second part of the book provide a more qualitative analysis of how capital cities carry out their governance and planning, allowing for a nuanced understanding of how they implement sustainability.

Including both quantitative and qualitative measures of sustainability policy and practice helps us avoid the problem that cities start at different points in terms of levels of environmental, economic, or equity performance given their geographic location, history, or development, and other factors that are often beyond the control of today's policy makers. By including the more qualitative analysis, we examine efforts that city stakeholders are making to achieve sustainable outcomes given their current conditions (Suter, Schaffner, Giddings, Orttung, & Streletskiy, 2017). The conclusion will tie together the quantitative and qualitative analyses to provide an assessment of whether capital cities are living up to their potential.

Independent variable

In choosing the hub as our independent variable, this book goes beyond the extensive theoretical literature arguing that the right combination of institutions, policy ideas, and networks makes some cities (or countries) more effective than others in achieving their goals (Taylor, 2016). In terms of the level of analysis, we are talking about the capital city's overall hub – the combination of the resources that capital cities bring together in ways that are not possible in other cities. The city-level approach is different from studies that focus on a single leader or firm as the driver of innovation. Capital city hubs are distinctive because they include the national government and embassies, attributes which are not found in non-capital cities. The other elements of the hub can be present in non-capital urban areas, but typically not to the same extent that they are available in national capital cities.

Who is the innovating agent? Who is driving and funding projects? Who is in the coalition that is promoting urban development? A common answer for this question is the claim that the innovation agent is the individual or the small firm, responding to market or government funding incentives. Under such conditions, the capital city or capital area region offers the institutional environment that should promote such individual or firm innovation because the capital promotes the exchange of ideas that leads to innovation. In contrast, this book argues that a broader institution – namely, the capital city hub – is the innovating agent.

Since the individual/firm and the capital city are at two different levels of analysis, the two hypotheses are not really alternatives. The broader, city-level approach gets at the question of under what circumstances do organizations learn and change, and there is already a substantial literature dealing with organizational (including governmental) change. This literature examines whether organizations change when they face external threats, when they have slack resources, or when organizational leadership changes. In terms of process, it examines whether they do so through some sort of organized search, through more randomly coming across ideas from the external environment, or through introducing new ideas through personnel turnover that brings new employees from the outside.

Is being a capital city a yes/no proposition, or are the various levels of "hubness" among capital cities? In fact, capital cities have elements of both properties. As cities, Washington, DC and Antananarivo may not seem to have much in common, but the capitals of the U.S. and Madagascar belong in a class that distinguishes them from other kinds of cities. However, the hub will likely function better in a place where there are greater economic resources, so the components of the Washington hub will undoubtedly be more robust than their Antananarivo counterparts.

However, by focusing on the capital city hub, we emphasize the various networks that make up the hub. This emphasis on the connections among people goes beyond earlier works about cities that focused more on the attributes of the cities than the mutual relationships among the various players that make up the hub (Sassen, 2001). Therefore, measuring the "hubness" of a capital city requires not just measuring the number and quality of universities, think tanks, tourist attractions, and other components, but the nature of the networks among the features of the hub.

A key element that capital cities provide in the functioning of the hub is the provision of the opportunity for members of the various networks to meet face to face. Campbell describes this feature of capital cities as particularly important for "the human element of special interest lobbying" and "the non-standardized process innovation of political work" (Campbell, 2003, p. 24). In fact, face-to-face interaction has been proven to improve people's ability to solve problems in all contexts, and is a key benefit of urban living (Cohen, 2018, p. 6; Glaeser, 2011, pp. 34–35). Studies of information-based services agree that they are dependent on face-to-face information exchange, and that the best place for these interactions are in highly developed central metropolitan cores (Hall, 2006). Even as other functions can be decentralized to other areas, the most information-rich activities remain concentrated downtown. In this sense, the development of information and communications technologies does not necessarily mean that face-to-face interactions no longer matter. Of course, personalized contact does not always lead to good outcomes, as Henry Hale shows in his wide-ranging examination of authoritarian "patronal" politics, in which corrupt personalized ties outweigh ideology to help dictators maintain their hold on power (Hale, 2014). Moreover, face-to-face discussion can be uncomfortable for many people, who might prefer to retreat to simply listening to media personalities whose opinions they already share (Berry & Sobieraj, 2014, p. 127).

The core argument of this book is that capital cities create a specific kind of knowledge that can only be fashioned in the physical space where all the elements of the hub are present. As Lundvall explains, "Learning by doing, using, and interacting (DUI learning) is a prerequisite for successful innovation and it results in tacit knowledge that is embodied in people and embedded in organizations that are rooted in specific localities (Lundvall, 2012)." Ultimately, what matters for developing innovative ideas is not knowledge of a specific topic, but an ability "to create and maintain contacts, dialogue and cooperation with others, who will later become suppliers, clients, workers and managers" (Asheim & Parrilli, 2012, p. 17).

In sum, we can measure the hubness of a city by assessing the level of connectivity among the key elements of the hub. As this introductory chapter explains, this book only makes a crude first approximation of such measurements, but it moves us closer to the ultimate goal of better understanding how the capital city hub functions in practice.

Alternative hypotheses

Does a well-functioning capital city hub necessarily lead to effective sustainability policy? Not everyone agrees that this proposition makes sense. Perhaps the most prominent alternative hypothesis is that capital cities do not innovate because they are too secure in the sense that they are sitting on a wealth of resources collected from around the country. Many citizens resent their capital because they view government workers as simply drawing large salaries funded by taxpayers without providing much in return (Cramer, 2016). Berlin, for example, is one city that has benefitted from its capital status during periods of financial crisis when federal spending supported local output (Slack & Chattopadhyay, 2009, p. 295). In contrast, non-capital cities innovate because they lack the security provided by ready access to the federal budget and face the necessity of coming up with new ideas to support themselves in the context of a competitive global marketplace. The literature on innovative cities rarely cites examples from capital cities. In fact, some authors have claimed that rustbelt cities are emerging hotspots of innovation, in part because they have no choice (van Agtmael & Bakker, 2017, p. 206). Along these lines, Taylor's explanation of why some countries are better than others at science and technology basically comes down to the idea that the innovators face an external threat which helps them overcome the opposition that any transformative invention encounters (Taylor, 2016).

The idea that the capital is unlikely to innovate because it is relatively secure raises the question of whether necessity is the only mother of invention. Or are people naturally creative even if they do not face an existential crisis? Swedes create more startups because they do not have to worry about losing their health benefits or sinking into poverty if they quit their jobs (Mounk, 2018, p. 232). This is a question at the core of American politics today: Should individuals be given basic levels of security on the grounds that they will innovate (most of us are makers), or should people be left face to face with disaster – no health insurance, the possibility of becoming homeless – so that they are required to innovate to survive? The latter approach relies on the assumption that most people are "takers" and would be happy to live off of government subsidies if given the chance. While this debate continues to rage in the U.S., European countries like Sweden claim that strong social safety net provisions allow for the kinds of risk-taking that spur innovation.

Another alternative approach is that a capital city will not generate new ideas because often the enemies of innovation are concentrated in the capital and can block progress. New ideas often threaten the status quo industries, and those firms usually have powerful representation in the capital that they can mobilize to block change. Innovation can undermine efforts to provide continuity, social order, and stability, and, accordingly, can be suspect (Juma, 2016). Different sectors of society have varying capacities to adapt, creating tensions that are ultimately resolved through the domestic political process concentrated in the capital. Such political processes can hinder the natural development of innovation for both new technologies and new policies.

An additional hypothesis is that capital cities are less likely to be centers of innovation because of the limitations placed on the local autonomy of these cities by national governments. Since these limitations vary substantially, this is at least potentially a testable hypothesis.

A final alternative hypothesis is that capital cities will ultimately lose out to global cities (Sassen, 2001). While capital cities once dominated international networks, they will play a shrinking role in the future because they are not as adept at working in the world economy and manipulating fast-moving financial flows as the global cities, and because they are rooted more in their country, which hinders their aspirations for global city status (Rossman, 2017, p. 256). Along these lines, the political state-centered status of the capital city might hinder its ability to play a role in the advancing international economy. Additionally, the success of globally important capital cities like London, Paris, and Tokyo might make them less effective as capitals, since many citizens of their countries resent their economic success. In short, the state-centered, hierarchical, conservative, bureaucratic, domestic-facing nature of capital cities might make them enemies to innovation, especially in contrast to the free-trading, dynamic network and flow oriented, outward-facing global cities of today and the future.

As this discussion makes clear, this book is based on normative assumptions about capital cities and sustainability. By arguing that the capital city hubs bring together the necessary resources required to make sustainability policies, we assume that the elements of the capital city hub combine in productive ways to promote sustainability policies that advance the cause of intergenerational justice. This assumption posits that the various players have a commitment to a constructive philosophy of social transformation based on using public policy to promote a productive balance between economic, equity, and environmental goals. As the case studies included in this volume show, however, these assumptions do not always hold. Some cities do not meet these standards for promoting sustainability, and in some cases, the hub function is not sufficiently well developed to achieve the predicted outcome. In the individual chapters, we will examine these assumptions more carefully, and then explore their successes and failures in the conclusion. That concluding analysis will show the strength of our hypothesis and that of the alternatives.

Key elements of the hub

As Figure 1.1 shows, the hub at the center of each capital city includes the national government, the capital city government, the research community (think tanks and universities), NGOs, embassies and the rest of the international community, corporations, media, and tourism. By definition, the focus of any capital city is the institutional and personnel resources of the national government. These include the elected officials of the executive and legislative branches, the judges of the judicial branch, and the staff personnel required to support them.

The U.S. capital provides a useful example of how this all works. In 2016, in Washington, DC, Virginia, and Maryland (the national capital area), there were 440,000 federal civilian employees, or about 21 percent of the 2,047,799 total U.S. federal civilian workforce (not including military personnel and post office employees).[4] While nearly four-fifths of civilian federal employees are based outside of the capital city – with large groups in California (141,158) and Texas (114,170) – no other part of the country has the same concentration of federal employees and expertise as does the capital.

The federal workforce includes well-educated and highly trained professionals who deal with every aspect of governance and the national economy. Since government rule-making affects everything from agriculture to finance and high-tech industries, the locus of power attracts all manner of lobbyists who hope to influence who gets what, when, and how, in Harold Lasswell's classic definition of politics (Lasswell, 1936).

The capital city government

The capital city government consists of the mayor/city manager, city council, and staff members who define and implement capital city policies. By interacting with the other elements of the hub, these city officials gain the benefits of their physical location, which gives them the possibility to develop more sustainable policies across the range of city concerns, from growing the local economy to ensuring high-quality living standards for the population.

The research community

Surrounding the federal workforce in any capital city is a large research community based in think tanks and universities. Globally, there are 7,815 think tanks, according to the University of Pennsylvania's 2017 Go To Think Tank Index (McGann, 2018, p. 6). The Index defines think tanks as "public-policy research analysis and engagement organizations that generate policy-oriented research, analysis, and advice on domestic and international issues, thereby enabling policymakers and the public to make informed decisions about public policy." Of these, 1,835 are in the U.S. and 1,770 are in Europe. Overall 90.5 percent of the think tanks were created since 1950. Nearly one-quarter of U.S. think tanks (397 institutions) are located in Washington, DC (McGann, 2018, p. 10). Think tanks come in a variety of forms, either autonomous, government-financed, political party affiliated, university affiliated, or corporate.

The U.S. capital also hosts at least five major research universities, which likewise cover a wide range of topics. These universities prioritize scientific discovery and the creation of new technology, drawing on the Prussian model first developed in 1810 (Taylor, 2016, p. 94). While other parts of the U.S. have developed extensive intellectual resources, including Boston, San Francisco/Silicon Valley, and Seattle, the capital area is unique for focusing specifically on policy and its implementation. While Washington's research universities are not in the same league as Harvard or Stanford, their social science and policy research is among the most prominent in the country. The collection of such resources in one place has been shown to stimulate collaborations and output (Hoekman, Frenken, & Oort, 2009; Mayer et al., 2016). Moreover, research demonstrates that universities encourage the presence of greater human capital and innovation (as measured by a boost in patenting) in the regions where they are located (Valero & Reenen, 2016). Similarly, the presence of universities raises gross domestic product (GDP) per capita.

Advocacy NGOs

Since capital cities make policies of all types, they naturally attract people who want to influence this policy in various ways. High-tech corporations such as Google and environmental organizations like the Sierra Club have both opened offices in Washington in order to shape government policies and form the national and international narrative in ways that support their causes. In 2016, total spending on lobbying was $3.15 billion, with 11,167 lobbyists working the halls of power, according to the Center for Responsive Politics.[5] These numbers fluctuate and were higher at the beginning of the Obama administration than at the beginning of the Trump era. The U.S. Chamber of Commerce, National Association of Realtors, and the pharmaceutical and health industry were the top spenders on lobbying activities.

Beyond the lobbyists, the capital hosts the national headquarters for numerous non-profit community groups that engage in a range of advocacy activities for the arts, community development, criminal justice, the environment, health, homelessness, LGBT issues, racial and ethnic concerns, gender, and senior citizens.[6] As Abbott pointed out in his history of Washington's evolution "the seat of government slowly attracted national institutions and organizations – many of which now locate in Washington because of each other's presence, not because of the city's character as a special and specific place." Moreover, he points out that the experts who run these organizations come from a "national pool of talent (Abbott, 1999, pp. 4–5)." Abbott worries that the people who make up the hub are somehow disconnected from Washington the evolving southern city, but nevertheless describes the fundamental hub feature of the city from its early origins:

> Washington was a networked city from the very start. Residents of the modern metropolis may be more self-conscious about their position in information space than were earlier generations, but 1820s congressmen in their little boardinghouse cliques and breakfast clubs were also "symbolic analysts," to use the 1990s term for brokers and traders in information. They energetically worked to build far-flung networks in support of abstract ideas. They were also as isolated from a regional Washington as are today's short-term association executives and jet-setting think tank gurus. For both, Washington has been an address rather than a community.
>
> (Abbott, 1999, p. 24)

Even if the association professionals are focused more on each other and their personal networks than the larger city, of particular relevance to this book's argument is that the capital city hosts numerous important professional city associations – the National League of Cities, U.S. Conference of Mayors, National Association of Counties, and the International City and County Management Association. An important part of what these groups do is disseminating knowledge about innovations and best practices to their member governments. The connection to that knowledge is, at least marginally, more readily available to members of the capital city government than it is to city governments elsewhere.

Embassies and the international community

Given the presence of embassies, capital cities are a natural link between the broader international community and domestic situation. As Gottmann has pointed out, a capital city cannot avoid serving as a hub for the interplay between these external networks and the internal economic and political structure of the country (Gottmann, 1977). Capitals are at the intersection between the global and local (Meisel, 1993, p. 3). The international community can serve as a source of sustainability ideas that can be adopted in capital cities and disseminated throughout the rest of the country. The capital can also be a gateway for the outflow of good ideas from the country into the wider international community. In terms of science and technology innovation, the ability to link into international networks is crucial in connecting domestic networks of STEM labor, entrepreneurs and investors with foreign markets for exports, investment capital, and sources of technical knowledge and skills (Taylor, 2016, p. 157). The kinds of social networks available in capital cities are much better at doing this than governments or markets by themselves. By bringing together the people and information quickly and easily, the networks reduce the costs and risks of innovation.

In Washington, the U.S. State Department under Hillary Clinton organized a Greening Embassies Forum on Earth Day 2009 to bring together the 175 diplomatic representations in the city to promote sustainability efforts across eight indicators. DC Mayor Vincent Gray joined the effort in April 2012, making the embassy program part of his plan to increase sustainability in the capital.[7] The Finnish embassy is the only LEED Platinum embassy in the city and provides a model in terms of practices that others can emulate. In partnership, the U.S. embassy in Helsinki is also LEED platinum, making them the only two embassies in the world that have achieved this level of sustainability.[8]

Many of the embassies based in Washington are focused on promoting products made in their country and by their companies, and many of these products can fit into the sustainability category. Embassies play this role around the world. The Dutch embassy in South Africa, for example, promoted sustainable business in the logistics sector, working to accelerate the adoption of advanced Dutch techniques by South African companies (Pross, 2010).

The international community is well represented in Washington, as it is in other capitals. In fact, the international organizations that have a city focus may be a lot more important in diffusing policy ideas than individual embassies. These include the German Marshall Fund, the Organisation for Economic Co-operation and Development (OECD), and International Monetary Fund (IMF), all of which have programs on cities and hold seminars and briefings that are open to the research community in the region. The agencies benefit from the presence of a specialized workforce of international experts based in the capital (Abbott, 1999, pp. 148–149).

Corporations

While corporations exist in every city, they play an important role in the capital city hub and its ability to generate sustainability policies. Businesses are the largest employer, source of financing, and greatest innovator in the world today, exceeding even the role of governments (Ehrenfeld, 2008, p. 197). Since today's environmental and climate change problems are so complex, government policy is necessary to develop effective solutions, but it is not sufficient by itself (Henderson, Gulati, & Tushman, 2015; Stavins, 2011). President Trump's decision to pull the U.S. out of the Paris accord on climate change, for example, has opened the door for corporations to take action.

While all corporations seek profit, they play different roles in terms of sustainability. Some corporations make their money selling outdated equipment or rely on fossil fuels that pollute the environment. By influencing policy to protect their market, they can block cities from achieving sustainable outcomes. In contrast, other corporations can make efforts to evolve their product line or emphasize new technologies that conserve resources or improve efficiencies in ways that help them derive profits while also promoting outcomes that are greener and more equitable (Ehrenfeld, 2008, pp. 201–204). Firms working in capital cities can contribute to sustainability policies by helping to drive industry self-regulation, developing common standards, triggering cultural shifts, and shaping understandings of what is appropriate (Henderson et al., 2015, p. 11).

The media

The national capital city naturally concentrates a large number of newsmakers, and many of the events that take place in the city are of interest to a large national and international audience. The result is a large press corps representing a variety of local, national, and international outlets that focuses on the federal government but also reports on other developments in the city, including sustainability issues. With such a large media presence, what happens in the capital city can easily be disseminated to a wide audience.

Tourism

The capital city tourist industry attracts a variety of tourists who come to the city for leisure, business, and educational purposes. Tourism is integral to the process of creating new national symbols, interpreting existing sites, and determining how the country should be presented to the rest of the world (Maitland, 2012). Beyond their capital city features, the cities also offer visitors a strong set of cultural assets including museums, monuments, cafes, bars, and restaurants, where people can meet and exchange ideas. (Diekmann & Cloquet, 2012).

The tourists who visit capital cities provide a network of contacts between the capital city and the rest of the country and world that many other cities do not possess. Ideas developed and promoted in the capital can be seen by the tourists and taken back to their hometowns for implementation there. Since the City Beautiful Movement from the late nineteenth and early twentieth centuries, the U.S. capital has been a "laboratory for comprehensive planning for transportation, parks, and core redevelopment" (Abbott, 1999, p. 178). Washington today, for example, can be seen as a model for its bike sharing system[9] – it was the first region in North America to experiment with third-generation bike sharing systems (Buehler & Stowe, 2016, p. 195). The built environment is also a paragon since the U.S. Green Building Council named Washington the first LEED Platinum city in the world.[10] The capital has more LEED-certified projects per capita than any state, and many of these are in the public schools. The citation noted Washington's role in reducing greenhouse emissions, clean energy innovations, and progress on other environmental metrics. Mayor Muriel Bowser boasted, "As the nation's capital, we have a special obligation to lead the way on environmental issues."[11] Similarly, Copenhagen can take pride in the Copenhagen City & Port Development Corporation which reimagined the port, an under-utilized part of town, to turn the city into one of Europe's richest, upending the situation from 30 years before in which it faced "17.5 percent unemployment, an outmigration of population, the loss of manufacturing, the decline of taxing capacity, and an annual budget deficit of $750 million" (Katz & Noring, 2017).

Linkages among the elements of the capital city hubs

For the capital city to play its hub function, the various members of the different groups described earlier must interact with each other in ways that generate new knowledge that leads to the promotion of sustainability policy and practices (Mayer & Cowell, 2014). The more comprehensive the thinking and interaction, the better (Webb et al., 2018). One common approach is to focus on the interaction between the government and the private sector. In the case of homeland security contracting, Mayer and Cowell argue that federal procurement for homeland security products acts as a powerful economic development force and that the DC region has solidified and expanded its importance as a central knowledge hub in America's gunbelt, where most of the country's arsenal is produced (Mayer & Cowell, 2014, p. 226). According to Mayer and Cowell (2014), the need for face-to-face meetings to work out contract details and other requirements for information exchange facilitated the emergence of a critical mass of contractors in the Washington DC region. Along these lines, a key component of what happens in capital cities that is different from regional innovation systems is that the government is often the main customer in purchasing products from companies and therefore shapes what they produce (Mayer et al., 2016, p. 13).

Despite these firm conclusions, Mayer and Cowell were not able to determine whether what happens in DC leads to innovations in the homeland security area or whether the firms "just fulfil specifications they co-develop with government agencies." They believe that answering this question tells us more about the nature of the economic milieu in the capital city and "whether the capital city as a knowledge hub is innovative and creative or whether it is highly protected and lethargic regarding innovation" (Mayer & Cowell, 2014, p. 243).

But the linkages among the various components of the hub go far beyond this conventional example and shape new policies and practices that affect sustainability issues as much as they do security concerns. Activists and practitioners are aware of the need to better conceptualize and promote these linkages. There is already clear evidence that they can pay off, as Sirianni points out:

> In the sustainable cities field, for instance, grassroots actors have challenged growth machines and highway projects since the 1960s, but eventually began to build much greater framing and policy capacity as they were joined by architects, planners, green builders, and transportation engineers who not only began to think in terms of sustainability, but whole professional associations also helped develop deliberative and information tools that enabled community groups to exercise voice and expertise.
>
> (Sirianni, 2017)

In a theoretical sense, capital cities promote what John Dewey had in mind in arguing that "the most comprehensive answer to a socially problematic situation must be thought of as the removal of barriers to free communication among the members of society so that problems can be solved in the most intelligent fashion" (Honneth, 2017, p. 60). "Potentials are unleashed and realized by the establishment of connections between hitherto isolated 'individual things,'" he wrote. The interaction among the different elements stimulates the generation of new ideas that can be tried out in the capital city and beyond. Examples of such collaboration are numerous, but one example comes from the Dutch Golden Age, when "the many thematic and stylistic connections that existed among Dutch masters of the latter part of the seventeenth century helped create a climate of excellence that propelled their genre paintings to levels of achievement rarely matched in the history of art" (Libby, 2017).

The case study chapters included in this book will test this Hub Theory by seeking to tease out the levels of connectivity among hub elements in the cities under study. In addition to estimating the strength of each of the nodes, the contributors will examine which of the different nodes on the hub interact with each other and how much interaction there is between these different nodes. The authors will ask what allows for greater connectivity and what hinders it. The chapters will examine to what extent the connectivity is spatial and takes place within the physical space of the capital, and to what extent it relies on the internet and other forms of communication. The more connections among the nodes, the stronger the hub is and the more likely it will be to produce effective sustainability policies.

Dissemination

Once the ideas are generated in the capital city, the question then becomes how they are disseminated to other audiences. There are three models that describe how the dissemination process works and each capital city could employ one or all three simultaneously. The models are:

1 the capital city innovates new ideas and then disseminates them at the local, national, or international levels,
2 the capital city collects good ideas from other cities and redistributes them to the rest of the country and world, or
3 the capital city demonstrates how to implement others' laws and innovations, especially those received from abroad.

Innovating new sustainability policies and practices, as in the first model, gives the capital city the most important role. While disseminating such ideas is not an important job for city officials, it can happen through mechanisms such as professional associations and the interaction of city officials at professional conferences. It is also possible that the capital city can come up with new ideas, but that these ideas are implemented in other areas, diminishing the geographical connection between initiator and innovation. Nevertheless, the ideas would not exist without the effort of those in the capital.

Figure 1.2 illustrates the second model, in which the capital city collects good ideas from other cities and helps a further group of cities to adopt them, as well. While the capital city may have the potential to innovate, it also is in a position to benefit from the myriad leading-edge ideas that bubble up as a result of grassroots efforts around the country. In fact, this model accepts as plausible the idea that capital cities are not the most likely innovator of sustainability ideas and that other cities outside the capital region play this role. In the case depicted in Figure 1.2, San Francisco was the first American city to ban plastic bags in retail stores, in 2007. Washington, DC adopted the practice in 2009, and subsequently

Figure 1.2 Plastic bag bans in the U.S. (2007–2017)

cities around the country imposed their own bans, as well. California and Hawaii imposed state-wide bans on the bags after the cities demonstrated that the practice was effective, while legislatures in Michigan, Missouri, Arizona, and Idaho took the opposite course, adopting state-wide laws that prevented cities from banning the bags. Evidence from California demonstrates that the ban is having a big impact on the local environment: According to a 2017 survey, "plastic bag litter had dropped by 72 percent when compared to 2010. Plastic bags now account for less than 1.5 percent of all litter, compared to nearly 10 percent in 2010."[12]

Rossman describes the third model, redistributing innovations from the international market to the domestic hinterlands, as the work of a capital city that cannot innovate on its own (Rossman, 2017, p. 257). He cites Buenos Aires as No. 20 on the Global Cities Index, but just No. 245 in the hierarchy of innovative cities.

Capital city contexts

The world's approximately 200 capital cities exist in a wide range of different geographic, economic, and natural contexts. The case studies included in this book come from different parts of the world, not only in geographic terms, but also regarding their levels of socio-economic development. Accordingly, their sustainability challenges, as well as the growth and governance issues they are facing, differ greatly. Table 1.1 employs the coarse measure of U.S. $ gross national income (GNI) per capita as a proxy to show where the various cities stand.

The case studies will show how these profound disparities affect our interpretative framework of the capital city hubs.

Beyond the different economic contexts in which the various capital cities exist, it is important to consider the variation of urban structure of a given country. In many less developed countries we see a primate city dominating the urban scene, with little economic energy or urban opportunities left for the rest of the country. Wolman's Chapter 2 in this volume addresses this situation more fully.

A last point to bring up is the differing relationship between capital cities and the natural environment where they are situated. Stockholm, for example, is particularly attuned to the natural environment. Generally speaking, however,

Table 1.1 U.S. $ GNI per capita of case study countries

USA	$58,270
Denmark	$55,220
Sweden	$52,590
Egypt	$3,010
Vietnam	$2,160
Cambodia	$1,230
Madagascar	$400

Source: World Bank Atlas, 2017 data

there is a correlation between levels of economic development and environmental externalities, as illustrated in the "environmental Kuznets curve."

As cities begin to develop, they have a negative impact on the environment. But once they reach a certain point, they have the resources required to reduce their negative impact. It is worth adding that this fits well with the picture of conventional environmental externalities: impacts from the lack of sanitation and solid waste removal services, and the presence of air pollution and industrial emissions tend to be mastered and put under control after a certain point of maturity of economic growth. Greenhouse gas emissions, on the contrary, tend to increase beyond the turning point for all sorts of reasons, and remain associated with economic growth, unless city leaders and other stakeholders are able to decouple the two through their policies.

Roadmap for the book

After this introductory chapter, the book has two additional overview chapters, followed by a series of case studies. Hal Wolman's Chapter 2, "Capital cities and regions: a general comparative overview" lays out the key reasons for studying capital cities, including their role and function and why they might be expected to be policy innovation generators. It describes how capital cities differ from non-capital cities with respect to population, economies, innovation generation potential, and local governance. It also describes how capital cities differ among themselves. While each capital city possesses a hub-based network of networks, capital cities differ from each other, inter alia, in terms of their political structure, status as primary or secondary capitals, and whether they are located in the developed or developing world. Capital cities differ from global cities because they focus on policy innovation rather than finance or trade development.

Stanislav E. Shmelev and Irina A. Shmeleva's Chapter 3, "Indicators for measuring the performance of smart and sustainable capital cities," identifies a set of indicators for measuring the performance of capital cities across a variety of sustainability and smart city categories. To do this, it examines the existing sets of indicators and evaluates them. Using a database of over 140 global cities, including both national capitals and others, the chapter builds a comprehensive CO_2 emissions model explaining 80 percent of the variance in urban CO_2 emissions across the globe. While a variety of factors matter, OECD capital status plays a special role in reducing per capita CO_2 emissions. The most plausible hypothesis explaining this outcome is the ground-breaking OECD work on urban pedestrianization and traffic management in the early 1970s. Capital cities are not the most prominent leaders in sustainability, with exceptional cities like San Francisco outperforming them. However, the national capitals of Stockholm and Seoul lead in terms of numerous policies and performance. This assessment lays the groundwork for measuring sustainability more systematically among capital cities and further testing hypotheses linking capital city status and sustainability.

The next section of the book includes a series of case studies from the developed and developing world. The studies highlight both cities that have been successful in grappling with sustainability policy and those that have failed, so that we would have variation across the dependent variable. We purposefully chose a wide range of cities and sustainability policies to demonstrate that the capital city hub can function across a variety of issue areas.

The first case study city provides a detailed analysis of what we mean by the capital city hub. Sahar Attia and Costis Toregas's Chapter 4, "Moving the capital to promote sustainability innovations: analyzing Cairo's future plans," explores the hub function of capital cities in the context of Egypt's capital Cairo. Egypt's most important city is a particularly interesting example because planners are in the process of moving key capital functions, financial institutions, innovation centers, and embassies to a new city to improve their effectiveness and sustainability. In contrast to the current congested and crowded metropolis, the new city will incorporate smart features with high levels of efficient mobility. The chapter illuminates the main theme of the book by examining whether it is possible to extract the essential hub functions from an existing city and transfer them elsewhere. It also includes a case study showing how universities and NGOs can work with residents of the city's massive informal developments to improve water accessibility and quality in the existing city of Cairo.

Anna Helm and Nicole Forsberg's discussion in Chapter 5, "Stockholm: where sustainability meets technology," focuses on Stockholm as a hub for CleanTech innovation and entrepreneurship. Stockholm is now the second most prolific tech hub in the world on a per capita basis. The chapter argues that the city's success derives from a symbiotic and mutually supportive relationship among government, industry, research centers and academia, as well as favorable relations with numerous civil society groups represented in the city. It shows how Stockholm's collaborative culture makes the hub function well. Stockholm represents an example of success for capital city innovation where there is extensive collaboration among the various actors.

Marta Andhov and Alexandra Andhov's Chapter 6, "Using public procurement to promote sustainability in Copenhagen," examines how the city of Copenhagen uses public procurement processes to promote sustainability goals. The authors, two lawyers based at the University of Copenhagen, show how public procurement policy and practices link various elements of the capital city hub and advance human rights, anti-corruption, labor movement, and environmental goals. Many of the procedures implemented in Copenhagen are derived from European Union (EU) procurement guidelines, and city officials counter efforts to water them down in coordination with NGOs, universities, media, and other hub components. All of these actions generate strong demand for sustainability policies from the public.

Linda J. Yarr and Nguyễn Ngọc Lý's Chapter 7 examines "Innovations in public participation for sustainability and resilience in Hanoi." As the historic and culturally significant capital of the Socialist Republic of Vietnam, a unitary state under single-party rule led by the Vietnam Communist Party, Hanoi offers a richly textured example of the hub effect on advancing policy and practice on sustainability. While the country as a whole undergoes a sometimes wrenching transition from war-torn poverty to middle-income status and heightened geopolitical responsibilities, Hanoi faces multiple challenges to its efforts to secure the well-being of its citizens amidst global warming, the temptations of untrammeled economic development, and an inadequate and overly complex public administration structure. With a focus on two case studies – the movements to preserve the city's tree canopy, and protection of its ponds and lakes – this chapter reveals how civil society actors leveraged Hanoi's resources as a hub to prompt city authorities to innovate with new mechanisms for public consultation on projects that had a powerful impact on the environment.

Jennifer Lessick's Chapter 8, "Delivering clean water in an African capital," develops an African case study. Antananarivo, Madagascar's capital and its political, economic, and cultural center, is home to over 1.3 million inhabitants, and growing. At a time when 92 percent of the country's population lives on less than $2 a day and less than 50 percent have access to clean water, the capital city has proven effective in sustainably increasing water accessibility for its inhabitants. Since 2010, the city of Antananarivo and the state-owned water supplier JIRAMA were able to provide over 700,000 residents of the capital with a reliable and efficient clean water supply, benefiting from the assistance of outside actors such as Water and Sanitation for the Urban Poor. Although the functioning of the hub system overall is extremely limited, a few elements of Antananarivo's capital city hub helped implement change in an otherwise slowly developing country. With one of the poorest populations in the world, Madagascar's government faces great challenges throughout the island not only in implementing sustainable innovation, but also in adapting to a changing climate. Given these conditions, it will be difficult to spread much of the gains achieved in the capital to the rest of the island.

Anthony Bigio's Chapter 9, "Phnom Penh, an unsustainable capital city," looks at Cambodia's unsustainable capital city Phnom Penh. The key question here is: Can economic growth, environmental protections, and social sustainability be reconciled despite poor governance, corruption, and a lack of rule of law? As in Hanoi, the main drivers for innovation are international agencies working with the national government and NGOs, but few of their ideas or plans have an impact on public policy. Key obstacles include poor governance, corruption, lack of rule of law, business activity controlled by the political elite, ongoing city growth with no real planning, insufficient energy supply, extreme congestion, and insufficient water supply, sanitation, solid waste and flood management. Phnom Penh provides an example of a city government that is not pursuing sustainability policies and where the hub is not functioning. Vested interests are strong and the government represses its opponents, making it difficult for the hub to function well. International agencies have developed a Phnom Penh Green City Strategic Plan and some neighborhood groups, especially those threatened by flooding, are interested in implementing it, but it is not clear if this relatively weak support base is enough to drive real change. Part of the problem is the dominance of investment from China, which does not include any requirements for sustainability.

Robert W. Orttung's concluding Chapter 10, "Conclusion: missed opportunities," synthesizes the main findings of the theoretical and empirical chapters. The chapter will argue that despite the achievements of some capital cities in specific areas, capital cities are not doing enough to take advantage of their hub functions. It will lay out areas for further research.

Notes

1 Except for hosting embassies, subnational capital cities in federal countries have all (or most) of the hub actors, albeit in smaller quantities. The presence of the hub actors there suggests that these cities also might be to some degree hubs of innovation, at least relative to non-capital cities. While beyond the scope of this book, such cities might be a fruitful topic of future research.
2 While local capital city governments are the focus of this book, capital city hubs can also produce and disseminate innovations through a host of other mechanisms, such as advocacy groups, universities, and the media, among others.
3 https://sustainabledevelopment.un.org/?menu=1300 (accessed November 28, 2017).
4 www.governing.com/gov-data/federal-employees-workforce-numbers-by-state.html (accessed November 27, 2017). DC employed 173,421, Virginia 136,377, and Maryland 130,402.
5 www.opensecrets.org/lobby/index.php (accessed November 28, 2017).
6 www.american.edu/ocl/volunteer/Nonprofit-Directories.cfm
7 www.washdiplomat.com/index.php?option=com_content&view=article&id=8296:embassies-pledge-to-go-green-as-part-of-dc-sustainability-push-&catid=1485:april-2012&Itemid=498 (accessed November 27, 2017).
8 Interview at the Finnish Embassy in Washington, DC on September 7, 2017.
9 www.gwhatchet.com/2017/09/13/d-c-to-test-bike-share-program-ditching-docking-stations/ (accessed November 28, 2017).
10 www.usgbc.org/articles/washington-dc-named-first-leed-platinum-city-world (accessed November 28, 2017).
11 http://wamu.org/story/17/08/31/d-c-named-worlds-first-leed-platinum-city/ (accessed on September 15, 2017).
12 www.mercurynews.com/2017/11/13/editorial-success-californias-first-in-the-nation-plastic-bag-ban-works/ (accessed November 13, 2017).

References

Abbott, C. (1999). *Political Terrain: Washington D.C. from Tidewater Town to Global Metropolis*. Chapel Hill: The University of North Carolina Press.

Asheim, B. T., & Parrilli, M. D. (2012). Introduction: Learning and Interaction – Dirvers for Innovation in Current Competitive Markets. In B. T. Asheim & M. D. Parrilli (Eds.), *Interactive Learning for Innovation: A Key Driver Within Clusters and Innovation Systems* (pp. 1–29). New York: Palgrave Macmillan.

Barber, B. R. (2013). *If Mayors Ruled the World: Dysfunctional Nations, Rising Cities*. New Haven: Yale University Press.

Berry, J. M., & Sobieraj, S. (2014). *The Outrage Industry: Political Opinion Media and the New Incivility*. New York: Oxford University Press.

Buehler, R., & Stowe, J. (2016). Bicycling in the Washington, DC Region: Trends in Ridership and Policies Since 1990. In D. Hyra & S. Prince (Eds.), *Capital Dilemma: Growth and Inequality in Washington, DC* (pp. 180–206). New York: Routledge.

Campbell, S. (2003). *The Enduring Importance of National Capital Cities in the Global Era*. University of Michigan Urban and Regional Research Collaborative Working Paper Series, URRC 03-08.

Caradonna, J. L. (2014). *Sustainability: A History*. New York: Oxford University Press.

Cohen, S. (2018). *The Sustainable City*. New York: Columbia University Press.

Cramer, K. J. (2016). *The Politics of Resentment: Rural Consciousness in Wisconsin and the Rise of Scott Walker*. Chicago: University of Chicago Press.

Diekmann, A., & Cloquet, I. (2012). How "Capital" Are Capital Cities on the Internet? *Current Issues in Tourism*, *15*(1–2), 19–33.

Ehrenfeld, J. R. (2008). *Sustainability by Design: A Subversive Strategy for Transforming our Consumer Culture*. New Haven: Yale University Press.

Glaeser, E. (2011). *Triumph of the City: How Our Greatest Invention Makes Us Richer, Smarter, Greener, Healthier, and Happier*. New York: The Penguin Press.

Gordon, D. L. A. (2006). Capital Cities in the Twentieth Century. In D. L. A. Gordon (Ed.), *Planning Twentieth Century Capital Cities*. Abingdon: Routledge.

Gottmann, J. (1977). The Role of Capital Cities. *Ekistics*, *44*(264), 240–243.

Hale, H. E. (2014). *Patronal Politics: Eurasian Regime Dynamics in Comparative Perspective*. New York: Cambridge University Press.

Hall, P. (2006). Seven Types of Capital Cities. In D. L. A. Gordon (Ed.), *Planning Twentieth Century Capital Cities*. Abingdon: Routledge.

Henderson, R., Gulati, R., & Tushman, M. (Eds.). (2015). *Leading Sustainable Change: An Organizational Perspective*. New York: Oxford University Press.

Hoekman, J., Frenken, K., & Oort, F. V. (2009). The Geography of Collaborative Knowledge Production in Europe. *Annals of Regional Science*, *43*, 721–738.

Honneth, A. (2017). *The Idea of Socialism: Towards a Renewal* (J. Ganahl, Trans.). Cambridge: Polity Press.

Juma, C. (2016). *Innovation and Its Enemies: Why People Resist New Technologies*. New York: Oxford University Press.

Katz, B., & Bradley, J. (2013). *The Metropolitan Revolution: How Cities and Metros Are Fixing Our Broken Politics and Fragile Economy*. Washington, DC: The Brookings Institution Press.

Katz, B., & Noring, L. (2017). *The Copenhagen City and Port Development Corporation: A Model for Regenerating Cities*. Washington, DC: The Brookings Institution Press.

Klemek, C. (2016). Exceptionalism and the National Capital in Late 20th-Century Paris and Washington, DC. In D. Hyra & S. Prince (Eds.), *Capital Dilemma: Growth and Inequality in Washington, DC*. London and New York: Routledge.

Lasswell, H. (1936). *Politics: Who Gets What, When, How*. New York: P. Smith.

Libby, A. (2017). *Vermeer and the Masters of Genre Painting (Exhibition Brochure)*. Washington, DC: National Gallery of Art.

Lundvall, B-A. (2012). Preface. In B. T. Sheim & M. D. Parrilli (Eds.), *Interactive Learning for Innovation: A Key Driver within Clusters and Innovation Systems* (pp. xi–xiii). New York: Palgrave Macmillan.

Maitland, R. (2012). Capitalness Is Contingent: Tourism and National Capitals in a Globalised World. *Current Issues in Tourism*, *15*(1–2), 3–17.

Matson, P., Clark, W. C., & Andersson, K. (2016). *Pursuing Sustainability: A Guide to the Science and Practice*. Princeton: Princeton University Press.

Mayer, H., & Cowell, M. (2014). Capital Cities as Knowledge Hubs: The Economic Geography of Homeland Security Contracting. In S. Conventz, B. Derudder, A. Thierstein, & F. Witlox (Eds.), *Hub Cities in the Knowledge Economy: Seaports, Airports, Brainports*. Farnham, GB: Ashgate.

Mayer, H., Sager, F., Kaufmann, D., & Warland, M. (2016). Capital City Dynamics: Linking Regional Innovation Systems, Locational Policies and Policy Regimes. *Cities*, *51*, 11–20.

McGann, J. G. (2018). *2017 Global Go to Think Tank Index Report*. Philadelphia: Think Tanks & Civil Societies Program, The Lauder Institute, The University of Pennsylvania.

Meisel, J. (1993). Capital Cities: What Is a Capital? In J. Taylor, J. G. Lengelle, & C. Andrew (Eds.), *Capital Cities Les Capitales: Perspectives Internationales International Perspectives* (pp. 3–5). Ottawa: Carlton University Press.

Mounk, Y. (2018). *The People vs. Democracy: Why Our Freedom Is in Danger and How to Save It*. Cambridge, MA: Harvard University Press.

Portney, K. E. (2013). *Taking Sustainable Cities Seriously: Economic Development, the Environment, and Quality of Life in American Cities*. Cambridge: The MIT Press.

Pross, E. M. (2010). *Sustainable Business and the Role of the Dutch Embassy*. Twente: University of Twente.

Rossman, V. (2017). *Capital Cities: Varieties and Patterns of Development and Relocation*. Abingdon: Routledge.

Sassen, S. (2001). *The Global City: New York, London, Tokyo, Second Edition*. Princeton: Princeton University Press.

Sirianni, C. (2017). Civic Innovation: Yesterday, Today, and Tomorrow. *Perspectives on Politics*, *15*(1), 122–128.

Slack, E., & Chattopadhyay, R. (2009). Comparative Conclusions. In E. Slack & R. Chattopadhyay (Eds.), *Finance and Governance of Capital Cities in Federal Systems* (pp. 292–326). Montreal and Kingston: McGill-Queen's University Press.

Stavins, R. N. (2011). The Problem of the Commons: Still Unsettled After 100 Years. *The American Economic Review*, *101*(1), 81–108.

Suter, L., Schaffner, C., Giddings, C., Orttung, R., & Streletskiy, D. (2017). Developing Metrics to Guide Sustainable Development of Arctic Cities: Progress & Challenges. In L. Heininen, H. Exner-Pirot, & J. Plouffe (Eds.), *Artic Yearbook 2017*. Akureyri: Arctic Research Forum and University of the Arctic.

Taylor, M. Z. (2016). *The Politics of Innovation: Why Some Countries Are Better Than Others at Science and Technology*. New York: Oxford University Press.

Valero, A., & Reenen, J. V. (2016). *The Economic Impact of Universities: Evidence from Across the Globe*. NBER Working Paper No. 22501.

van Agtmael, A., & Bakker, F. (2017). *The Smartest Places on Earth: Why Rustbelts Are the Emerging Hotspots of Global Innovation*. New York: PublicAffairs.

Webb, R., Bai, X., Smith, M. S., Costanza, R., Griggs, D., Moglia, M., . . . Thomson, G. (2018). Sustainable Urban Systems: Co-Design and Framing for Transformation. *Ambio*, *47*, 57–77.

World Commission on Environment and Development. (1987). *Our Common Future* (Brundtland Report). www.un-documents.net/our-common-future.pdf

2 Capital cities and regions

A general comparative overview

Harold (Hal) Wolman

We begin this chapter by addressing the question that underlies our entire enterprise: Why does it make any more sense to study capital cities than to study cities whose name begins with "P" or whose name consists of three or more syllables? Our focus is on the unique set of network relationships in the capital city environment that constitute a hub, which, we hypothesize, ought to result in the generation of innovative local public policies. We begin with a discussion of the role and function of capitals, followed by a description of how capital cities differ from one another, focusing particularly on capital city political form. This is followed by a presentation of how capital cities differ from non-capital cities with respect to population, economies, innovation generation potential, and local governance. We present both theory and results from empirical research on why and the extent to which such differences occur. We conclude by setting forth concerns that each chapter will address.

Why study capital cities?

The fundamental reason for studying capital cities is that they are different from other cities in important ways. This book argues that the existence of a capital hub should give capital cities the ability to innovate sustainability policy and practices. Within this context, it is important to understand how capital cities differ from other cities in their countries and from global cities generally. In addition, comparing capital cities to other cities can lead to a better understanding of cities in general. Just as studying the institutions and social and political behavior in a foreign country can throw light on features of one's own country that had previously been unexamined or accepted as natural, so can focusing on a national capital highlight features that illuminate the other cities in the system.

Capital cities differ from other cities particularly with respect to population characteristics such as occupational structure and education – capital residents are higher skilled and better educated. As a consequence, since these characteristics are associated with growth, capital cities tend to have higher levels and rates of growth of population, income, and employment than do other cities in the system. They also differ from other cities in terms of their economic base and industrial structure. They are likely to be disproportionately knowledge and information cities rather than industrial cities, with a greater share of workers in government, professional and business services, and high-tech sectors, and fewer in the manufacturing sector.

Capital cities also are characterized by a nexus of institutions that exist to be in close proximity to government. These include professional and business associations, lobby groups, non-profit advocacy groups, research institutions (including national research laboratories and centers), universities, foreign embassies, international organizations and donors, domestic and international media, and intergovernmental organizations. In addition, capital city regions generate private contractors who establish themselves near the capital in order to gain advantage as part of the national government procurement process. Given these characteristics, it is the expectation of this book that capital cities have the potential to be major centers of regional innovation systems (RISs). Whether they actually are major innovators will depend, at least partly, on the nature of capital city local governance and whether local government institutions have the autonomy and capacity to engage in and implement innovative policy.

Research on capital cities

Regardless of how the "Why study capital cities?" question is answered by researchers, there exists a literature on capital cities, although Mayer et al. (2018) characterize it as relatively sparse and lacking a coherent set of theories. The literature deals with a variety of concerns including capital city roles and functions, economy, symbolic meaning, architecture, political structure, local governance, and the balancing of national and local concerns. The literature tends to focus more on capital cities in federal countries than in unitary systems.[1]

The capital cities included as chapters of this book are Antananarivo, Madagascar; Cairo, Egypt; Copenhagen, Denmark; Hanoi, Viet Nam; Phnom Penh, Cambodia; Stockholm, Sweden; and, to some extent, Washington, DC, USA. Data for these cities on a variety of topics discussed in this chapter are presented in Table 2.1.

Capital cities: their role and functions

At the most basic level, the role of the capital city is to house the institutions of the national government, usually including legislative, executive (including the official residences of the elected leader and head of state), and judicial branches.[2] The capital city is also the locale through which the country conducts relations with foreign countries, and therefore is likely to house embassies and a range of international institutions.

Beyond that, the capital city serves a variety of symbolic functions both domestically and internationally. For example, Rossman (2017: 13) contends that:

> The main task of capital cities is to help the nation to visualize itself and to present the nation to the rest of the world. With this in mind, capitals

Table 2.1 Selected data for capital city case studies

Country	*Capital*	*Country GDP per capita, 2015 ($)*	*Capital city population, 2010*	*% change capital city population, 2000–2010*	*Capital city region per capita household income, 2010 ($)*	*National per capita household income, 2010*	*Capital city region advanced services as a % of total capital city regional employment, 2010*	*Country advanced services as a % of total national employment, 2010*	*Capital city region manufacturing as a % of total capital city regional employment, 2010*	*Patents per 10,000 residents of capital city region, 2015*	*Patents per 10,000 residents of nation, 2015*
Denmark	Copenhagen	44,524	539,542	8.09	27,562	25,308	58.7	48	6.30	3.76	2.28
Sweden	Stockholm	44,323	847,073	12.89	33,304	29,182	56.70	50.60	6.90	5.68	3.26
United States	Washington, DC	51,911	601,723	5.19	58,866	42,267	55	52.90	1.50	1.39	1.43
OECD Median		36,818	767,773	5.95	28,055	25,815	48.70	36.74	7.06	1.36	0.97
Cambodia	Phnom Penh	1,163	1,663,341	38.83	525	213	27		47		
Egypt	Cairo	2,413	16,899,000	19.36	10,000				18.6		
Madagascar	Antananarivo	1,469	1,300,000								
Viet Nam	Hanoi	6,046	6,900,000	10.10	1,923	1,387	47.98	33.80	31.80		

Source: OECD Regional Statistics, World Bank, World Population Review, Populationof2017.com, CEICdata.com, CAPMAS, Egypt's Ministry of Planning, Monitoring, and Administrative Reform, and Wikipedia

Note: Empty boxes denote data available

> represent the ideal image of the country and country's history, a miniaturized version of the nation, so to speak. Because of this representation function capitals are sometimes perceived as the equivalent to the nation as a whole.

As a consequence, the capital city houses museums, monuments, and archives related to the history of the nation and memorializing that history. It also frequently contains art and cultural institutions that reflect the nation's history and the grandeur of the nation.

Given these roles, capital cities inevitably take on an ancillary role: the provision of security for the institutions of national government and foreign embassies, both in terms of the physical protection of buildings and the personal safety of officials. This role often also includes the protection of the related national museums and institutions and extends to the surrounding areas of these buildings and to the transportation system connecting them to each other and to the residences of government officials.

Finally, the capital city's role includes that of local governance for the residents who live within the city's boundaries. For most countries, this role is provided through formal and viable institutions of local government. However, in some cases, it is functionally provided by the national government. Local governance includes providing local services for and imposing taxes on local residents and businesses, albeit with varying degrees of local service responsibilities, revenue raising authority, and autonomy. It also includes engaging in inter-local interactions, negotiations, and collaboration with other units of subnational governments in the region or bordering it, and with higher levels of government.

With respect to the latter, the role of the capital city as local government and its role as the seat of the national government often conflict and result in a struggle to balance the national and the local interest. As Rowat (1973: xi) observes, the national government "wants to control and develop the capital in the interests of the nation as a whole, while the people of the capital naturally wish to govern themselves to the greatest extent possible." Some of this tension is resolved through constitutional provision, some through national legislation, and some from informal consultation and politics that occur between the capital city local government and national government.

Differences among capital cities

Capital cities differ substantially among themselves. In his comprehensive survey of the literature on capital cities, Scott Campbell (2003: 6–7) lists 15 characteristics on which there are differences among capital cities.[3] Following, we focus on differences in capital city institutional structure, including form of national government (unitary vs. federal), and the economic level of the country, because these two features are most likely to affect the ability of the capital city to develop innovative policies. The literature suggests that city-state-type capitals in relatively rich countries are most likely to innovate.

Formal capital city institutional structure

Building upon Rowat (1973), Wolman et al. (2007) classify the capital cities in their 11-country study as:

- Capital cities that are federal districts or territories within a federal system (Washington, DC, Brasilia, Canberra, Delhi, Mexico City).
- Capital cities that are city-states within a federal system (Berlin, Vienna, Brussels).
- Capital cities that are located as a local government unit *within* a state, province or other subnational level of government (Ottawa, Bern).
- Capital cities that are a city within a unitary system (London, Paris, Copenhagen, Stockholm).

There is clearly some relationship between type of form and other characteristics of the country and its history. Slack and Chattopadhyay (2009: 303) observe that unique capital districts tend to be newer planned cities in federal systems. In European countries, particularly in unitary systems, the capital is usually the largest and most important city and has attained its capital status through historical circumstances (London, Paris, Rome, Madrid, Stockholm, Copenhagen, Oslo, Warsaw). In federal systems, however, as Rossman (2017) points out, the capital city is usually not the largest city in order to prevent dominance of one entity among the system's units.

There has also been a debate about the normative question of which form is most desirable and under what circumstances. Rowat (1973) and Slack and Chattopadhyay (2009) argue that the district model is particularly appropriate for countries with regional divisions or with diverse populations where the diversity is reflected in the geographic concentration of the various groups. Slack and Chattopadhyay (2009: 298) contend that:

> One of the main justifications for the federal-district model is to ensure that the national set of government is not under the jurisdiction or the control of any of the states or provinces. In this way, the model reduces the risk of favoritism among different regions in the country and eliminates potential friction between the federal and the state governments in overlapping areas of jurisdiction.

In the same manner, this arrangement reduces potential conflict among ethnic groups by avoiding placing the capital within a subnational polity dominated by one of the groups, thus providing a politically neutral and acceptable solution (see Belgium, for example). Rowat (1968: 349) also observes that, "It is often argued that, as a matter of principle, the capital of the federation should not be governed by the laws of any one state in the federation." A unique capital district solves this problem; at the same time, however, there is the danger that such a district will have too much control over local residents, with the potential for depriving them of political rights enjoyed by other citizens of the nation.

The "city-state" model, as Nagel (2013) notes, reduces the possibility of "asymmetrical" citizenship by providing residents of the capital with the same political rights as those possessed by members of other states. However, as Rowat (1968) observes, it reduces the nation's ability to exert political control over its own capital by providing a local government with its own constitutional powers. In addition, Nagel cites the potential for conflict between the city-state capital and other states in the federation that do not possess the special relationships that capital status brings.

The capital as a city within a member state of a federation or as a city within a unitary system places the capital's residents in a position similar to all other residents of the nation, but it reduces the ability of the nation to affect the actions of the capital city, though less so than is the case with city-states, since the latter are endowed with constitutional powers (Slack and Chattopadhyay, 2009). As a result, Slack and Chattopadhyay (2009: 302) contend that, "What differentiates this model from the other two models is that the provincial or state government has more control over municipal affairs in the capital city than does the federal government." They note, for example that in both Bern and Ottawa, the federal government has no institutional means of intervening in city policy decisions, and all federal-capital city communication occurs through the canton/province. Finally, when there is a diverse population, the nation risks political and social tension if it places the capital in a city that reflects that diversity. However, the additional control over its own politics and policies for these capital cities often is accompanied by corresponding problems. As Slack and Chattopadhyay (2009: 303) note:

> Although cities in a province/state experience less federal intervention compared to those under the federal district model, they often complain that the federal government does not grant them adequate recognition or financial support to meet the cost of their function as the nation's capital. Their funding depends more on the province or state government than on the federal government.

Despite the four-fold typology discussed earlier, the structure of each capital city has unique characteristics, and many capital cities appear to have features of two or more forms. As Slack and Chattopadhyay note quoting Graham Sansom, for example, Canberra (the Australian capital territory), despite being set up as a unique capital district, "is positioned and structured as a city-state within the federation and the territory government is treated in a very similar way to the states and Northern Territory, apart from the special arrangements for capital-city planning" (Slack and Chattopadhyay, 2009: 300).

With respect to the innovation generating capacity, it would seem, a priori, that, holding all else constant, independent districts would have less potential, particularly because they are most likely to have less autonomous local governments, while city-states would have the greatest capacity, since they are likely to have the most elaborated and dense set of institutions and network relationships.

In addition to the political structure, any effort to compare capital cities must take into account the level of economic development of the capital city's country. Level of economic development may be associated with the potential for capital city innovation generation through a variety of paths, including the degree of connectivity of the network information, communication, and transport systems, and the capacity of institutions of local government, particularly with respect to professionalism, corruption, the rule of law, and the availability of adequate fiscal resources. See Table 2.1 for the relative economic strength of the countries discussed in this volume.

Differences between capital cities and other cities in their country

Capital cities are likely to differ from other cities in a myriad of ways. In an overarching view of the ways in which capital cities are unique, the geographer Jean Gottmann (1983: 88) argues that:

> The capital function secures strong and lasting centrality; it calls for a special hosting environment to provide what is required for the safe and efficient performance of the functions of government and decision-making characteristic of the place. . . . That hosting environment is rather complex; the capital will tend to create for and around the seats of power a certain kind of built environment, singularly endowed, for instance, with monumentality, stressing status and ritual, a trait that will increase with duration. The capital's social environment will also be very special due to the congregation there, ex officio, of certain categories of workers: civil servants, diplomats, lobbyists, newsmen and, of course, politicians, who are not to be found in similar numbers in other than capital cities. . . . As a result of the capital function, a large and diversified aggregate of labor develops, in the quaternary category of occupations. The political personnel normally form only a part of that labor force. Economic and cultural functions ancillary to the government activities are important components and may come to form a majority in the capital.

Jane Jacobs has a less favorable view of capital city "uniqueness." As Jacobs (1985: 231) puts it:

> [C]apital cities thrive on transactions of decline. When a city's principal function is being a capital . . . it is obvious that the more transfer payments, subsidies, grants, military contracts and promotion of international advanced-backward trade, the greater the work and the prosperity in the capital city. . . . Behind its busyness at ruling, a capital city of a nation or an empire, vivacious to the last, at length reveals itself as being a surprisingly inert, backward and pitiable place.

These distinctions provide a variety of reasons and theories for why capital cities and their regions might be expected to have larger and more rapidly growing populations and economies than other cities in their countries. Population and economic size and growth are obviously closely related, but for the sake of clarity we next address each separately (or as separately as is possible).

Population

Theory and expectations

There are theoretical reasons to expect that capital cities will have greater population than other cities – or rather, will have greater population than they otherwise would have if they were not capitals. First, capital cities and their regions are places of residence for elected politicians and non-elected policy makers and bureaucrats who are permanent workers for the government (or who occupy permanent positions that will continue to have occupants even if present incumbents leave as a result of election results and administrative changes). The existence of the government will attract other ancillary activities such as lobby groups, professional associations, researchers, etc., who will also reside in the area (Gottman, 1983; Zimmermann, 2010; Wrede, 2016; Mayer et al., 2016; Rossman, 2017). In addition, businesses or business people who wish to interact with the government through contracting or research and development or to influence government policy and regulatory decisions will be more likely to locate in the capital city or region, and to reside there. As governments in nearly all countries have expanded over time, frequently even in the face of economic downturns and reductions in government budgets (which often turn out to be reductions in the *rate of growth* of government budgets), population growth of capitals has increased and is likely to continue to do so (Carroll and Meyer, 1982).

In addition, capitals are likely to contain additional characteristics that lead to both population and economic growth. As Wrede (2016: 120) notes, "Capital cities have many advantages because their function as a capital requires certain unique features. To provide easy access for domestic and foreign officials, major nodes of rail and road networks and major airports must be closely located." Turner and Turner (2011) contend that governments tend to place more and better-quality infrastructure in the capital, as a commitment to the capital's symbolic representation of the country to both domestic and foreign visitors. The same holds true for museums, monuments, parks, and many other amenities. And, since government officials reside in the area, they are able to express their own preferences for good local infrastructure and services and higher amenities. Given that civil servants and political officials in most countries are likely to be more highly educated and receive higher pay than the average citizen, the capital is likely to have many restaurants and entertainment and recreation opportunities that are consistent with the preferences of these sectors of the population (Zimmermann, 2010). And these characteristics make the region a desirable place to live, even for people who are not connected to the government.

Furthermore, to the extent that these infrastructure and amenity items are funded at least partly, if not mostly, by government revenues provided by taxpayers nationally rather than locally, the price of these goods to the local citizenry is reduced, and consequently the quantity of such goods purchased increases. The result is that infrastructure and amenities in capital cities are likely to exceed those in other cities, thus increasing the area's attraction as a place to live (see Wrede, 2016; Dascher, 2000).

Ades and Glaeser (1995) provide an addendum, theorizing that developing and authoritarian country capitals are likely to have disproportionate population and population growth (they also find empirical support for the proposition). Their argument is directed at primate cities, but – given that most capital cities are, in fact, primate cities – should apply to developed democratic country capitals as well. They posit that concentration of population in a large capital isolates the government from possible instability and violence in the hinterlands, provides it with large rents to buy off potential opposition, and allows it to extract disproportionate resources from the hinterlands.

Empirical research

There is a body of research that indeed supports the expectations that capital cities and their regions have a larger population than they otherwise would if they did not contain the capital and that these cities grow more rapidly than do other cities. Capital cities are typically one of the largest, if not the largest, cities in their countries, making them primate cities. Ades and Glaeser (1995) examined the largest city in each of 85 countries and found that the city was 42% larger if it were also the national capital.[4] Of the 35 OECD countries, the median capital city population was 768,000 (in 2010), and the capital city was the primate city in the country in 29 of them. In only two of the OECD countries did the capital city rank below the fifth most populous city in the country: the United States (Washington, DC, which ranked 10th) and Australia (Canberra, which ranked eighth). Kim and Law (2014) examined population size in the United States and Canadian national capitals (Washington, DC and Ottawa, respectively) and found that:

> Political capital status had a positive and statistically significant effect on population after controlling for a variety of factors that are likely to be correlated with economic activity. . . . For the national capitals, our estimates suggest that Washington DC attracted 68% more in population due to its status as the nation's capital in 1900 but its impact grew over six-fold to 447% by 2000. For Ottawa, its capital city effect on population agglomeration was around 159% in 1920 and rose to 589% in 2000.

(See also Carroll and Meyer, 1982) Primate capitals also prevail in the developing world. Rossman (2017: 234) examines a much larger set of countries including both developed and developing countries and finds that, "The majority of political capitals (83 percent) are located in the biggest cities of their respective countries: only in 25 out of 146 relatively big countries the capital is not sited in its largest city."

In 28 of the OECD countries, the capital city population increased from 2000–2010, and in 18 of these, it increased at a faster rate than the population of the country. The capital region's population also increased in all but three countries, and the suburban population (the non-city portion of the region) increase exceeded that of the city in 27 cases.

Economies

Theory and expectations

The previous discussion on population has already suggested some of the theoretical rationale for why capital city economies might have higher incomes and more rapidly growing economies than other cities, ceteris paribus. The capital region economy is a good example of Keynesian export-led growth (Dascher, 2000; Heider et al., 2016; Zimmermann, 2010; Wrede, 2016). A part – often, a very large part – of the capital city export base consists of government services. Government employees in the capital provide government services which are, in effect, exported to the remainder of the country and paid for through national taxes. Organizations ancillary to the government such as lobby groups,[5] professional associations, research institutes, and foreign embassies also locate in the capital region, and they, too, have employees who are likely to have above-average incomes. These employees reside in the region and their income recirculates in the area in conformance with the Keynesian multiplier effect (i.e., that portion of income which is not saved by residents or used to import goods and services from outside of the region is, instead, spent on goods and services within the region, and the additional income thus derived then repeats the same process). Since government employee earnings are likely to be higher than the national average, this multiplier process produces more vibrant secondary effects than in the typical non-capital city region. In addition, the government purchases inputs through the procurement process to support its operations and services, and many of these inputs are likely to come from local sources as a means of minimizing transport costs. Furthermore, as Carroll and Meyer (1982) theorize, government in the modern age has tended to expand relative to the rest of the economy, providing an export base with built-in long-term growth and even relative protection from economic volatility.

Like all large city regions, capital cities benefit from agglomeration economies, and the larger the city and region, the greater the agglomeration economies. Given the propensity of research and development and high-tech firms to locate in the capital region in order to have access to government decision makers and contracts, knowledge spillovers in these sectors are likely to result in the creation of new firms through spin-offs from existing firms. To the extent that these new firms (or, indeed, existing firms that locate in the area to gain access to the national government) then sell some of their product outside of the capital region, these products become part of the region's export base and contribute additional income, with consequent multiplier effects to the area economy, as described earlier.

In addition, the capital city stands as a symbol of the nation – and is thus likely to have national monuments, museums, buildings, and parks that serve as major attractions to tourists, both foreign and domestic. Thus, tourism is also likely to be an important part of the region's export base. And, as Zimmermann (2010) notes, these capital related functions can, by definition, "be given to only one region and then constitute a big bonus for this region and a relative disadvantage to the rest" (See also Mayer et al., 2016).

Other characteristics of the capital city not directly related to its capital functions are also likely to give the region an advantage as a location for economic activity. Carroll and Meyer (1982) and Turner and Turner (2011) contend that national governments will likely ensure that capital regions are infrastructure rich, particularly with respect to transportation infrastructure and facilities (airports, railway stations and connections), as well as information technology infrastructure. While Carroll and Meyer (1982: 572) suggest that this is done out of "ideological commitment to the importance of the state as well as community self-interest," the result is that such infrastructure redounds as well to the benefit of other economic activity not directly related to capital city functions.

Empirical research

Research results support the expectation that capital cities and their regions experience higher rates of employment and income growth than do similarly situated cities and regions that are not capitals, or than would have been the case were they not capital cities. Data from OECD countries indicate that capital city regions have higher levels of economic output per capita and higher disposable income per capita than are the case nationally. While capital regions in OECD countries accounted for only 21.9% of the national population in 2010, they produced 27.2% of the national gross product. With respect to income, residents of the capital region had higher per capita disposable income in 2010 than the nation as a whole in 25 of the 29 OECD countries for which such data are available, with the median being 19% higher.

Carroll and Meyer (1982) employ a variety of regression models to estimate the effect of U.S. capital city status on income growth over time. They find (p. 572) that:

> In each equation estimated, the effects of capital city status are positive and large. They are not always significant, but the results seem quite regular. Across each time period from 1949 to 1974, both state and national capitals grow more in income than do other cities.

Turner and Turner (2011) conducted research on the effect of capital status on income growth in the United States and found that capital city income growth was more elastic with respect to state expenditure than was the case with non-capital cities in the state. Dascher (2000) compared West Germany county towns that had lost their county town status in reforms undertaken during the 1960s and early 1970s to those that had retained their county seat status. Controlling for a host of other factors, he found that those that had retained their county seat status had employment gains seven percentage points higher from 1970 to 1987 than did those towns that had lost their status.

Innovation generation potential

In this book, we set forth the proposition that capital cities are likely to generate more innovative activity than other cities, and that this will be particularly manifested in local government policy, including sustainability policy. The logic of the argument follows.

A characteristic of cities is the agglomeration economies that they generate. Agglomeration economies are savings[6] that result from firms locating in geographic proximity to each other. They come in two forms, cost savings and "knowledge spillovers." Cost savings result from lower input costs (as a result, for example, of lower transport costs if suppliers have enough of a market to locate in the region, lower search costs for employees, and the availability of a wide range of skills in the labor force) and increased productivity that reflects external benefits to firms resulting from their proximity to each other. These are external economies of scale, i.e., they are savings that accrue to firms simply by virtue of their location.

The second, and more important for our purposes, form of agglomeration economies consists of "knowledge spillovers." As Gordon and McCann (2000: 520, discussed in Wolman and Hincapie, 2014: 136, *emphasis in original*) observe, these are:

> informal networks of individuals *across* firms (and also across other related institutions such as trade associations, universities, research institutes, and labor organizations) [that] result in the transmission of tacit knowledge that leads to innovation and the adoption of advanced and improved techniques related to production processes, marketing, and research. These networks are based on interpersonal relations and trust and are said to embody social capital that is embedded in them.

The relevant "firms in capital cities" for our purposes are the actors and entities that populate the hub in the diagram in Chapter 1 and constitute an interactive network. The most relevant actors are the national government, the local government of the capital city, foreign embassies, international organizations and donors, universities and research institutes, national NGOs, professional and business associations, lobby groups, and the national and international media.

The actual nature of the networks and their effectiveness in generating innovative activity will vary from capital to capital. Characteristics of local governance for the capital city will, in particular, affect the extent to which the capital city will generate sustainability policy innovations.

Indeed, Mayer et al. argue (Mayer et al., 2016) that these characteristics make capital cities and their regions particularly good locales for generating innovation. They observe (Mayer et al., 2016: 207) that:

> The economic geography literature, for example, conceptualizes capital cities as "information cities" (Castells, 1989), "national information brokers" (Abbott, 1999, 2005) or "transactional cities" (Gottmann, 1977) where complex relationships between government, private sector and third sector actors form a distinctive economic system which is spatially manifested through their interactions, which in turn produce information and knowledge that are important to their nation (Abbott, 1999; Feldman, 2001; Gerhard, 2007; Markusen, Hall, Campbell, & Deitrick, 1991). In this sense, we can consider capital cities as RISs.

The extent to which capital cities are "knowledge cities" is captured in their industrial and occupational structure relative to the rest of the nation. Capital city regions in the OECD countries are characterized by a much greater share of employment and output in the advanced services (information and communications; finance and insurance; real estate; professional, scientific and technical services; and public administration) than is the nation as a whole. The median value of these advanced services as a percentage of total employment in 2010 was 48.9% in capital city regions, compared to 36.5% in the nation's overall economy. On the other hand, the median value of manufacturing as a percentage of total employment was only 7.1% in national capital regions, compared to 13.7% in the national economy. Capital city regions are centers of the post-industrial globalized economy.

Mayer et al. (2016: 206ff) contend that capital cities, as a result of knowledge interactions between the public and private sectors, develop "a unique regional innovation system (RIS)." The key driver of this interaction is the public procurement process. They contend (2018: 7) that, "Through public procurement, capital cities become crucial places where both sectors meet and firms interact with other firms and government agencies." Quoting Edquist and Hommen (2000: 5), Mayer et al. (2016: 13) argue that the procurement process fosters innovation when national government organizations "place orders for a product, service, good or system that does not yet exist, but which could (probably) be developed within a reasonable period of time." (For an in-depth discussion of public procurement for sustainability, see, in particular, Chapter 6 in this volume on Copenhagen.) Mayer et al. also argue that public procurement for innovation (PPI) serves to reduce the risks of innovation through providing cost savings from economies of scale as a result of high levels of demand and by taking on the role of first users giving the innovation a guaranteed market thus reducing transaction costs. They conclude that, "PPI in this sense can function as a very specific driver of a RIS and these processes need to be considered when examining capital city economies (Mayer et al., 2016: 13)."

These characteristics have led some of the theorists in the global city literature to point to the prominence of capital cities, particularly in the developed world, as global cities. These global cities serve as the nexus or hub of a globalized economic system that, it has been argued, has superseded individual nation-states as centers of economic activity and capital flow (see, for example, the essays in Knight and Gappert, 1989; also Sassen, 1991; Scott, 2001; Salet, Thornley, and Kreukels, 2003; and Hall and Pain, 2006). Indeed, 14 of the 25 top global cities in A.T. Kearney's Global Cities Report (A.T. Kearney, 2017) are national capitals.[7]

However, capital cities have government-related characteristics that distinguish them from non-capital global cities. Their population includes actors who embody a high degree of professional and technical knowledge related to policy issues, including sustainability policy: employees of the national government, universities, and research institutes with public policy expertise; professional and business associations; lobby groups on a wide variety of issues; national NGOs; and international organizations. We should expect, as a consequence, that policy innovations would be more likely to occur in capital cities than in non-capital global cities. OECD data does show that in 10 of the 16 countries for which data are available, overall innovation – as measured by patents per 10,000 residents – is higher in capital city regions than in the country as a whole.

The following chapters of this book provide a further initial test of this hypothesis.

Capital city governance: translating potential into policy

Where both the national and city governments support sustainability policies, there is often little conflict between the two and they work harmoniously. The case of Stockholm (Chapter 5 in this volume) demonstrates such a situation well. Where the national government does not support or prioritize sustainability policies and the city government does, as is the case typically in Washington, DC, when Republicans occupy the White House and a majority in Congress, the ability of the federal government to limit city actions matters greatly for the city's ability to move ahead with innovative sustainability policies. In these cases, the autonomy of the capital city matters.

Capital cities serve as local governments for those who live within their borders, and it is typically these local governments that must translate the resources of the hub into effective policies. Harris (1995: 244) observes that, in democratic polities, "The essential challenge in governing the capital city is to strike a balance between national and local interests." Capital cities must somehow structure a set of arrangements that protect the national interest in the capital, but also reflect the operations of local government and the preferences of local residents. How these two interests intersect and are integrated is a concern nearly all capital cities must address.

Young (2008) provides a stark example of these tensions as faced by the District of Columbia, the capital of the United States: "On the one hand, the federal government provides the city and region with a near recession-proof industry, an amazing array of cultural amenities, such as the Smithsonian museums, and the up-close spectacle of the world's most powerful government at work. On the other hand, the federal government's presence imposed tremendous functional and fiscal burdens on the District. The United States government also dramatically limits local governmental autonomy and even restricts the rights and powers of District citizens to receive representation and seek redress before the federal government." The constant tension between these two pulls may limit the effectiveness of the city government.

Van Wynsberghe (2003) arrays national capitals of 17 federal countries[8] along a dimension from those tilted towards national control to those tilted towards local control. The rankings from most national influence to least are: Abuja (Nigeria); Washington, DC; a group consisting of Asian capitals (New Delhi, Islamabad, Kuala Lumpur); a Latin American group (Caracas, Mexico City, Buenos Aires); Canberra; Ottawa; Brussels; a Germanic group (Berlin, Vienna, Bern); Belgrade; and Moscow. Harris (1995) concludes that none of the federal districts he examined provides the "full gamut" of self-government to its local citizenry available to other citizens and that "even in the most advanced democratic systems studied in this report – Canberra and Brasilia – the national governments have 'exclusive jurisdiction' and ultimate power over their federal districts (Harris et al., 1997: 89)."

In developed countries, the local governance of capital cities occurs through democratically elected institutions of local government, although in some developing nations, local governance – at least in terms of policy making – is functionally provided by the central government. However, even in the former cases, the need to balance national and local interests may result in local governments that differ in important ways from other local governments in the country. These differences may relate to local autonomy, fiscal autonomy, service responsibilities and revenue sources, fiscal assistance (particularly the extent to which the capital city is reimbursed for expenditures that support national government functions or for national government property that is exempt from local government taxation), and local government capacity. In each of these cases, the differences are likely to serve as limitations on local government, and thus raise questions about its ability – despite the favorable innovative hub network environment – to engage in innovative policies in sustainability or other areas.

The limitations on capital city local governance make comparison of capital cities across countries difficult, as well. A major problem in comparing the relationship of capital cities to their national governments across countries is sorting out differences that are due to the capital city–national government relationship from those that are due to differences in the nature of the entire system of intergovernmental relations between any two countries.[9]

Degree of local autonomy

The most basic question is whether the capital city has local self-government at all. Wolman et al. (2007) found that all 11 of the capitals in their sample had local self-government elected by their residents, although for some, local self-government through popular election had come relatively recently (Washington, DC in 1973; Brussels in 1989; Canberra in 1989; Brasilia in 1990; Mexico City in 1996).

But how much local autonomy do capital cities have relative to other cities in the country? Wolman et al. found that, as of 2007, several capital cities were subject to provisions whereby the national government could override local laws or refer contentious issues to joint national government-local government committees to negotiate differences.[10] In Washington, DC, the United States Congress has the ability to override local laws, including the budget and budget items, by a majority vote. In Australia, unlike for other state and local governments, the national government has the power to limit Canberra's power to legislate and to override laws that it passes.[11] In Mexico City and Delhi, the capital city must have its budget approved by the federal government (Slack and Chattopadhyay, 2009)

In some countries, formal or informal committees exist to resolve disputes between national and local interests. In Brussels, for example, constitutionally, the laws of the national government have precedence over the laws of the Brussels capital region and the federal government has the right to intervene only when the national interest is threatened. Any such potential conflicts are to be resolved through arbitration at the Council of State level. The federal government may also suggest that issues of urban planning, transportation, and public works that are of concern to it be sent through a "Cooperation Committee" consisting of four Brussels ministers and four federal government ministers for discussion and resolution (Van Wynsberghe, 2003). In Germany, the relationship between the federal government and its capital, Berlin, is set out in a Cooperation Agreement agreed to in 1992 when the capital was moved from Bonn to Berlin after reunification. The agreement calls for the creation of a joint committee for resolving any questions between the capital and the national government. Slack and Chattopadhyay (2009: 302) write that:

> In Bern, for example, there is a yearly meeting between the federal, cantonal, and city governments to discuss concerns. . . . There are also coordination schemes at the technical level, such as regular contacts for federal, cantonal, and city officials responsible for security and for cultural programming.

As these examples suggest, it appears that discussion – and sometimes, resolution of disputes involving national and local interests – frequently take place in informal or formal committees involving both parties. It appears, however, that in most of the mentioned cases, these mechanisms rarely – if ever – go beyond the stage of consultations.

Fiscal autonomy

With the exceptions of Mexico City and Washington, DC (see earlier), all the capitals among the 11 studied in Wolman et al. (2007) have local fiscal autonomy (i.e., they can make their own decisions about taxes and expenditures without approval of the national government) to the same extent as do other local (or state, in the case of city-states) governments. In Mexico City, the city budget must be approved by the national legislature (although there are several capitals where laws adverse to the national interest can, at least in principle, be overturned by the national government). Slack and Chattopadhyay (2009) note that among the 11 capital cities in federal countries examined in their edited volume, three (Abuja, Delhi, and Mexico City) must have their budget approved by the federal government.

Service responsibilities and local revenue sources

Slack and Chattopadhyay (2009) note that while the services for which national capitals are responsible vary greatly across countries, in most cases these service responsibilities are the same as for other city governments (or state governments with respect to capitals that are city-states) throughout the country. There are some exceptions, usually in capitals that are districts. Slack and Chattopadhyay (2009) also report that federal capitals on the whole receive municipal funding from the same sources as do other cities in the country.

Fiscal assistance from national government

Wolman et al. (2007: 14) conclude that among the 11 non-United States capitals they studied, "there is enormous variation in the extent to which capital cities are treated differently than other cities in terms of fiscal transfers, payments for capital city functions, and compensation for property exempt from taxation." Young (2008) reviews three major comparative capital studies[12] that examine, in total, 14 capitals, and found that seven (Berlin, Brasilia, Brussels, Canberra, Caracas, Ottawa, and Paris) of them receive some form of compensatory financial aid to make up for the added burdens of being the site of the national capital and for revenue lost as a result of tax exempt national government property.

Wolman et al. also note that even when capital cities do receive such compensatory aid, the amount which they receive tends to be relatively small in terms of their total budget. They also point out an important difference between the circumstances of most of these foreign capitals and that of Washington, DC, a capital city that since 1999[13] has received no compensatory assistance for its capital role. Unlike the United States, most of the national governments in their sample operate major programs that provide "equalization grants" to cities that are fiscally stressed – they take into account the fiscal capacity and needs of cities in providing these funds. In some cases (e.g. Germany and Sweden), there is a system of horizontal transfers among states or local governments; in these cases, the capital city/state sometimes benefits (e.g. Berlin) and sometimes serves as a net contributor (e.g. Stockholm).

Several cities receive compensatory assistance for their functions as a capital city. The Australian national government's financial relationship with Canberra is the same as that of the other states and territories in the Commonwealth, except with regard "to the special circumstances arising from the existence of the national capital and the seat of [the Australian government] in the Territory" (Australian National Government, 1988). Berlin also receives compensation for its financial burden as the capital city, including security expenditures, infrastructure needed for purposes related to being the capital city, and for cultural activities. In Brazil, a "Constitutional Fund for the Federal District" was created in 2002 to provide Brasilia with fiscal assistance; the revenues received must be used for public safety, education, and health. The French government provides funding to the capital, Paris, for equipment and operations in the city and for specific projects (Boyd and Fauntroy, 2002). Ottawa receives payment in lieu of taxes from the Canadian federal government for federal property that is exempt from local taxes (Harris et al., 1997). Wolman et al. (2007) note that Mexico is also supposed to compensate its capital, Mexico City, for tax-exempt property, but it appears to have done so inconsistently. While large capital cities like London, Stockholm, and Vienna may not be formally compensated for their status as a capital city, they may well be favored informally through funding arrangements that reflect the fact that they are capitals.

Local government capacity

Local governance effectiveness requires that local government possess the *capacity* to govern. Beyond legal and fiscal autonomy capacity includes adequate fiscal resources; employees chosen by merit rather than patronage and who possess policy knowledge and administrative skills; political and policy leadership that has its sources of authority in democratic legitimacy; functioning institutions of local governance, including a culture of professionalism; and strong and enforceable legal standards and cultural values against corruption. Local governments that are low with respect to capacity may not be able to take advantage of the favorable innovation generating hub environment for policy innovation.

While it is unlikely that capital cities will differ from non-capital cities in the same country on most of these capacity dimensions, differences among capital cities across countries may be substantial. Efforts to measure and rank local government capacity across countries are still in the conceptual stage. However, there have been many attempts to rank governance by countries, most based on surveys of perceptions. The World Bank's Worldwide Governance Indicators ranks countries on the "perceptions of the extent to which public power is exercised for private gain." For the countries included in our book, the two Scandinavian countries, Denmark and Sweden, both rank in the 98th or above percentile (with higher percentiles meaning less perceived corruption), while the

United States falls just below the 90th percentile. Viet Nam falls just above the 40th percentile, Egypt at the 31st percentile, and Cambodia just above the 10th.

Conclusion

In this chapter, we have examined how capital cities differ from each other and from other non-capital cities in their countries. In so doing, we have painted a picture of the capital city as a unique hub of institutions and actors that provide it the potential for generating and implementing innovative policies in sustainability and other areas. The composition and relative influence of these institutions and actors, as well as their integration into the resulting network, is likely to vary not only across capital cities, but also from one policy area to another. In addition, the potential for generating local policy innovations will also depend on local governance arrangements in each country, and particularly on the degree of autonomy provided local government and local government capacity.

These concerns will be addressed in each of the following country-specific chapters. Chapter authors will examine the institutions that populate the circumference of the hub presented in Chapter 1 and describe their relative influence in policy making, both in general and with respect to the specific policy with which the chapter is concerned. They will also discuss local governance arrangements of the capital city, including both the degree of local autonomy and local capacity available for innovative policy making. Each chapter will describe an innovative sustainability policy measure, what the measure does, how it was adopted, and the main network actors that influenced its adoption. When possible, each will also discuss the effectiveness of the measure and the extent to which it has been disseminated to other cities in the country.

Notes

1 The chapter relies heavily on secondary research, although some sections do include primary research. Much of the material is drawn from several publications that compare capital cities (Harris et al., 1997; Boyd and Fauntroy, 2002; Van Wynsberghe, 2003; Wolman et al., 2007; Young, 2008; Slack and Chattopadhyay, 2009; Nagel, 2013). Since these publications are not current – and some are more than 20 years old – the details may no longer be accurate. One way of interpreting these data is to view them, not as completely accurate in every respect at the present time, but as a picture of general tendencies and as providing both generalizations about capital cities as a group and also a sense of the range of variation among them.

2 Although in a few cases major government institutions may be split between two cities, each of which has capital functions (e.g. The Hague and Amsterdam in the Netherlands, and Berlin and Bonn in Germany). Slack and Chattopadhyay (2009) observe that three of the 11 countries in their edited volume – Canada, Nigeria, and Switzerland – have located the legislature in the capital but have decentralized at least some other institutions of government across the country.

3 1) population of the capital city; 2) population of the nation; 3) geographic size of the nation; 4) form of national government (e.g. democratic, authoritarian, etc.); 5) degree of centralization of the national state (including, presumably, federal vs. unitary systems); 6) era in which the capital city was created (age of capital city); 7) continuous

capitals vs. relocated capitals; 8) economic prosperity of the nation; 9) whether the capital city is the dominant city in the nation or the secondary one; 10) the economic base of the capital city; 11) unitary vs. split capitals (i.e., capital functions in two or more places) 12) geographic location of the capital within the country (central vs. periphery); 13) centrality of the capital with respect to population centers and transportation networks; 14) capital of monolingual vs. multilingual nation; and 15) capital as historical religious center.

4 Using only land size as a control, Galiani and Kim (2008: 5–6) found that national capital status in Latin America "increases [the city's] population by 232% as compared to 175% and 154% for U.S. and Canada respectively; for provincial capital status, the Latin American figures are 120% as compared to only 32% for the U.S. state capitals but somewhat comparably higher than 102% for Canada."

5 As Wrede (2016: 125) notes, "When the federal government has large rents to dispense, the number of lobbyists in the capital is also large. . . . The total amount of rent granted by the federal government can be expected to increase with the size of the budget."

6 Although these can be partially offset by congestion costs.

7 In order of ranking: London (2), Paris (3), Tokyo (4), Hong Kong (5), Singapore (6), Beijing (9), Washington, DC (10), Brussels (11), Seoul (12), Madrid (13), Berlin (14), Moscow (18), Vienna (20), and Amsterdam (22). New York City is the first ranked.

8 Argentina, Australia, Austria, Belgium, Brazil, Canada, India, Malaysia, Mexico, United States, Nigeria, Pakistan, Russia, Switzerland, Venezuela, and Yugoslavia.

9 For example, in many countries, cities – including the capital city – are provided aid by the national government to compensate for differences in fiscal capacity that might place cities at a disadvantage. Thus, many capital cities may appear fiscally better off than Washington, DC, but this is not because of a determination to respond to the special needs of the capital; it is because of differences in the way in which all cities in these countries are treated relative to cities in the United States.

10 Boyd and Fauntroy (2002) found that, as of 2002, in 5 of the 11 capitals they studied – Washington, DC; Caracas; Mexico City; Canberra; and Rome – the national government had the authority to override local decisions.

11 Although Harris et al. (1997) observe that no one whom they interviewed for their report remembered any case of the Australian federal government overruling a decision by the Australian Capital Authority (Canberra).

12 Wolman et al., 2007; Boyd and Fauntroy, 2002; Harris et al., 1997.

13 Prior to the Revitalization Act of 1997, the District of Columbia received a federal payment, which reached $660 million in 1996.

References

Ades, Alberto F., and Edward L. Glaeser. 1995. "Trade and Circuses: Explaining Urban Giants." *The Quarterly Journal of Economics*. 110 (1): 195–227.

AT Kearney. 2017. *Global Cities 2017: Leaders in a World of Disruptive Innovation*. Accessed August 1, 2017 at www.atkearney.com/documents/10192/12610750/Global+Cities+2017+-+Leaders+in+a+World+of+Disruptive+Innovation.pdf/c00b71dd-18ab-4d6b-8ae6-526e380d6cc4.

Boyd, Eugene P., and Michael K. Fauntroy. 2002. *Washington, D.C., and 10 Other National Capitals: Selected Aspects of Governmental Structure*. Washington, DC: Congressional Research Service.

Campbell, Scott. 2003. *The Enduring Importance of National Capital Cities in the Global Era*. Urban and Regional Research Collaborative, University of Michigan, Working Paper Series, paper 03–08.

Carroll, Glen, and John Meyer. 1982. "Capital Cities in the American Urban System: The Impact of State Expansion." *American Journal of Sociology*. 88 (3): 565–578.

Dascher, Kristof. 2000. "Are Politics and Geography Related? Evidence from a Cross-Section of Capital Cities." *Public Choice*. 105: 373–392.

Galiani, Sebastian, and Sukkoo Kim. 2008. *The Law of the Primate City in the Americas*. Working Paper, Washington University of St. Louis.

Gordon, I. R., and McCann, P. 2000. "Industrial Clusters: Complexes, Agglomeration and/or Social Networks?" *Urban Studies*. 37 (3): 513–532.

Gottmann, Jean. 1983. "Capital Cities." *Ekistics*. 88–93.

Hall, Peter Geoffrey, and Kathy Pain, eds. 2006. *The Polycentric Metropolis: Learning from Mega-City Regions in Europe*. London: Routledge.

Harris, Charles W. 1995. *Congress and the Governance of the Nation's Capital: The Conflict of Federal and Local Interests*. Washington, DC: Georgetown University Press.

Harris, Charles W., Hilary Cairns, Jeff Carpenter, and Andrew Stettner. 1997. *Foreign Capital City Governance: Representation, Governmental Structure, Finances, and Intergovernmental Relations in Six Capital Cities* (Presentation to the District of Columbia Financial Responsibility and Management Assistance Authority). Washington, DC: The D.C. Community Policy Forum and the Georgetown Public Policy Institute.

Heider, Bastian, Albrecht Kauffmann, and Martin T. W. Rosenfeld. 2016. "Does Administrative Growth Status Matter for Urban Growth? Evidence from Present and Former Capitals in East Germany." Halle Institute for Economic Research.

Jacobs Jane. 1985. *Cities and the Wealth of Nations: Principles of Economic Life*. New York: Random House.

Kim, Sukkoo, and Marc Law. 2014. "Political Centralization, Federalism, and Urban Development: Evidence from US and Canadian Cities." *Social Science History*. 40 (1): 121–146.

Knight, Richard V., and Gary Gappert, eds. 1989. "Cities in a Global Society." *Urban Affairs Annual Reviews*. 35.

Mayer, Heike, Fritz Sager, David Kaufman, and Martin Warland. 2016. "Capital City Dynamics: Linking Regional Innovation Systems, Locational Policies and Policy Regimes." *Cities*. 50: 206–215.

Mayer, Heike, Fritz Sager, David Kaufman, and Martin Warland. 2018. *The Political Economy of Capital Cities*. Abingdon: Routledge.

Nagel, Klaus-Jurgen. 2013. "Introduction: Asymmetry and the Capital." In Klaus-Jurgen Nagel (ed.). *The Problem of the Capital City: New Research on Federal Capitals and Their Territory*. Barcelona: Institut d'Estudis Autonomics.

Rossman, Vadim. 2017. *Capital Cities: Varieties and Patterns of Development and Relocation*. London: Routledge.

Rowat, Donald. 1968. "The Problems of Governing Federal Capitals." *Canadian Journal of Political Science*. 1 (3): 345–356.

Rowat, Donald. 1973. *The Government of Federal Capitals*. Toronto: University of Toronto Press.

Salet, Willem G. M., Andy Thornley, and Anton Kreukels, eds. 2003. *Metropolitan Governance and Spatial Planning: Comparative Case Studies of European City-Regions*. London: Taylor & Francis.

Sassen, Saskia. 1991. *The Global City*. New York, London, Tokyo, Princeton: Princeton University Press.

Scott, Allen. 2001. *Global City Regions: Trends, Theory, Policy*. Oxford: Oxford University Press.

Slack, Enid, and Rupak Chattopadhyay. 2009. "Comparative Conclusions." In Enid Slack and Rupak Chattopadhyay (eds.). *Finance and Governance of Capital Cities in Federal Systems*. Montreal: McGill-Queen's University Press: 292–326.

Turner, Sidney, and Richard Turner. 2011. "Capital Cities: A Special Case in Urban Development." *Annals of Regional Science*. 46: 19–35.

Van Wynsberghe, Caroline. 2003. "Federal Capitals: A Comparison." *Revue Internationale de Politique Compare*. 10 (1): 63–77.

Wolman, Harold, Jan Chadwick, Anna Karruz, Julia Friedman, and Garry Young. 2007. *Capital Cities and Their National Governments: Washington, D.C. in Comparative Perspective*. Washington, DC: George Washington Institute of Public Policy Working paper no. 30.

Wolman, Harold and Diana Hincapie. 2015. "Clusters and Cluster-Based Development Policy." *Economic Development Quarterly*. 29 (2): 135–149.

Wrede, Matthias. 2016. "The Political Economy of Capital City Size." In D. Meyer and F. Stübben (ed.). *Wirtschaftswissenschaft und Wissenschaftsdiplomatie*. Bamberg: BERG Public Economics Series, Vol. 28: 143–160.

Young, Garry. 2008. "Fiscal Relationships Between Capital Cities and Their National Governments." In *Building the Best Capital City in the World: A Report by DC Appleseed and Our Nation's Capital*. Washington, DC: DC Appleseed and Our Nation's Capital.

Zimmermann, Horst. 2010. "Do Different Types of Capital Cities Make a Difference for Economic Dynamism?" *Environment and Planning C: Government and Policy*. 28: 761–767.

3 Indicators for measuring the performance of smart and sustainable capital cities

Stanislav E. Shmelev and Irina A. Shmeleva

How can we measure the sustainability of capital cities? Do they do a better job of promoting sustainability policies and practices than other cities? There is a large and growing literature on urban sustainability indicators and how best to apply them. This chapter evaluates that literature and seeks to apply it to a set of global and capital cities. The overall finding is that the capital cities that can also be ranked as global cities do not differ much from global cities in general regardless of whether they are capitals or not. Interestingly, OECD capital cities rank high on the measure of walkability and use of public transportation, which is strongly correlated with lower CO_2 emissions. This finding connects back to the hub feature of capital cities discussed in Chapter 1 of this volume. Face-to-face contact is important for the capital city hubs to work effectively. However, in general, the finding that global capital cities do not perform much better than global cities as a whole suggests that capital cities are not fully utilizing their capacities to achieve sustainability outcomes.

The topic of urban sustainability has begun to receive a lot of attention in the EU, U.S. and increasingly China and Latin America since the Rio Summit of 1992, the Rio+20 Summit of 2012 and, especially, in light of the recent Habitat III forum held in Quito, Ecuador in 2016. The UNEP Green Economy Report highlighted urban sustainability as one of its important dimensions (UNEP, 2011). Moreover, the new UN Habitat World Cities Report firmly links the New Urban Agenda with the 17 UN Sustainable Development Goals (SDG) adopted in Paris at the end of 2015 (UN Habitat, 2016). In particular, SDG 11 "Sustainable Cities and Communities" aims to "make cities and human settlements inclusive, safe, resilient and sustainable" (UN, 2015b). The United Nations Economic Commission for Europe (UNECE) and the International Telecommunication Union (ITU) launched a new United for Smart and Sustainable Cities initiative in 2016.

Urban sustainability is defined as the multidimensional capacity of a city to operate successfully in economic, social and environmental domains simultaneously (Girardet, 1993, 2004, 2014; Naess, 1995; Hall and Pfeiffer, 2000; Bithas and Christofakis, 2006; Shmelev and Shmeleva, 2009; Dassen et al., 2013; Hall, 2014; Martin and Rice, 2014). According to Hall and Pfeiffer (2000), the taxonomy of cities is complex and cities cannot be classified on one dimension alone. The multidimensional nature of an urban system defines the central analytical approach for the sustainability assessment of cities used in this chapter, namely the methodology of Multi-Criteria Decision Aid (Roy, 1996), following an approach outlined in Shmelev (2017).

The Rome declaration adopted at the UN Forum on "Shaping smarter and more sustainable cities: striving for sustainable development goals" in May 2016 declared that "cities need to become smarter, with technological solutions deployed to address a wide range of common urban challenges" of sustainable development (UNECE and ITU, 2016). The EU's European Economic and Social Committee considers smart sustainable cities to be a tremendous source of growth, productivity and employment. A smart sustainable city, according to UNECE, is an innovative city that uses information and communication technologies and other means to improve quality of life, efficiency of urban operation and services, and competitiveness, while ensuring that it meets the needs of present and future generations with respect to economic, social and environmental, as well as cultural, aspects (UN ECOSOC, 2015).

One simple and straightforward method for measuring the sustainability of cities is to measure their carbon emissions. Cities contribute 75% of global CO_2 emissions, and will experience tangible effects from its consequences. Bai et al. (2018) assert that by 2030, millions of people and US$4 trillion of assets will be at risk from climate change-induced extreme events.

This chapter is organized as follows. The next section discusses data and the indicators used. Subsequently, we present the results of a regression analysis utilizing a database of 140 global cities linking urban CO_2 emissions with various external, infrastructure, policy, behaviour and technology variables. We then examine how capital and non-capital global cities perform, and develop three case studies. The conclusion examines what urban sustainability indicators tell us about the performance of capital cities.

Indicators for smart sustainable cities

Existing smart and sustainable cities indicator frameworks include the United Nations Guidelines and Methodologies on Sustainable Development Indicators (UN, 2007), EU Sustainable Development Indicators (EC, 2009), UNECE Sustainable Development Indicators Frameworks (UNECE, 2013), ISO 37120 standards on Sustainable Development of Communities issued in 2014 (ISO, 2014) and revised in 2018, the UN Sustainable Development Goals framework (UN, 2015b), and the UN Economic and Social Council Smart Sustainable City Indicator Framework (UN ECOSOC, 2015). Recently, there has been a growth of interest in indicator-based sustainability assessments for cities (Shen and Zhou, 2014; Mori and Yamashita, 2015; Wong, 2015; Yigitcanlar et al., 2015; Wei et al., 2015, 2016). These frameworks are discussed extensively in a range of comparative reviews (Valentin & Spangenberg, 2000; Spangenberg, 2002a, 2002b, 2005, 2017; Kierstead and Leach, 2008; Monfaredzadeh and Berardi, 2015; Hara et al., 2016; Manitiu and Pedrini, 2016; Ahvenniemi et al., 2017;

Garcia-Fuentes et al., 2017; Girardi and Temporelli, 2017; Klopp and Petretta, 2017; Pierce et al., 2017).

Our comparative analysis of the three most prominent assessment frameworks – UN SDG indicators, ISO 37120 Sustainable Development of Communities and UNECE-ITU Smart Sustainable City Indicators – identified a difference in focus; the balance between economic, social and environment dimensions; and some inconsistencies. The UN SDG indicator framework emphasizes the problems of developing countries. Its 249 indicators are often defined in an imprecise way, and thus are difficult to implement. The 2014 ISO 37120 standard includes more precisely defined indicators, though social and environmental aspects receive slightly greater prominence at the expense of economic and smart indicators. In contrast, the UNECE-ITU Smart Sustainable Cities Indicator framework is more balanced among the different dimensions of sustainability, is formulated with great clarity and benefits from a forward-looking strategic vision.

The selection of individual indicators for cities in this chapter draws on an earlier sustainable cities framework (Shmelev and Shmeleva, 2009), inspired by our dynamic sustainability assessments carried out for countries (Shmelev, 2011) and adapted for the urban scale Shmelev (2017). The process of indicator selection for the study was performed in two parts. First, a large set of criteria was analyzed, including economic indicators (income per capita at PPP, number of large companies headquartered in the city, creative industries employment), environmental indicators (CO_2 emissions per capita, share of nuclear energy, PM_{10} emissions, water use per capita, waste generation per capita, recycling rates) and socio-cultural indicators (unemployment rate, Gini Index of income inequality, life expectancy). After performing a Principal Component Analysis (Shmelev, 2017), identifying redundant variables and adding relevant dimensions, the set of criteria took its final shape numbering 20 criteria as a result of several iterations. Here we focus on CO_2 emissions as the most useful indicator.

The Environment Europe™ Sustainable Cities Database, a proprietary product which we have developed as our main data repository, includes 143 cities in Europe, North America, South America, Africa, Asia and Oceania. Our study draws on a wide range of sources from Eurostat (2016), city governments, UN Habitat, World Bank, the Carbon Disclosure Project (CDP) and Bloomberg. We illustrate the geographical spread of the database with CO_2 emissions data (Figure 3.1). CO_2 emissions are particularly high in Melbourne and Sydney, Dubai and Doha, Miami, Shanghai, Almaty and much lower in Stockholm, San Francisco, New York, Tokyo, Rio de Janeiro, Paris and Madrid.

Does capital city status matter for sustainability?

In this section we examine whether capital city status matters for reducing levels of CO_2 emissions, which we use as a proxy for sustainability. We test a variety of causes for such emissions using a cross-section regression analysis. The hypotheses we test were derived from assertions in the UN Guidelines on Sustainable Development Indicators (UN, 2007), which emphasized the interdisciplinary

Figure 3.1 CO_2 emissions

Source: Environment Europe™ Sustainable Cities Database, http://environmenteurope.org/, 143 global cities, 2018

connections between sustainable development indicators. The exact formulation of the hypotheses is based on our previous research outlined in Shmelev and Shmeleva (2009), Shmelev (2017) and Shmelev and Speck (2018). What becomes clear in this analysis is that capital city status is by far not the most important driver for CO_2 emissions.

The confirmation of our hypothesis of a highly significant correlation between the amount of CO_2 emissions and the share of coal, the most carbon-intensive technology at present in the energy mix (Figure 3.2), reinforces the need for an urgent transformation and decarbonization of the energy sector. Such cities as Sydney, Warsaw, Hong Kong, Denver, Portland, Los Angeles, Washington and Shenzhen have above-average levels of coal in the energy mix and exhibit high per capita CO_2 emission. On the other hand, such cities as Sao Paulo, Rio de

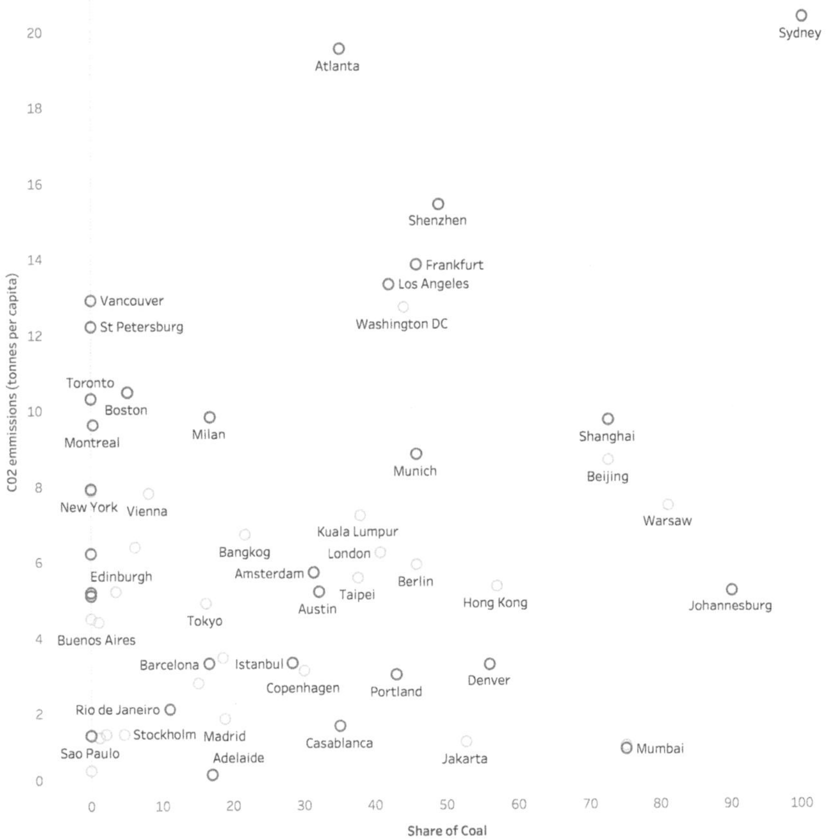

Figure 3.2 Correlation between CO_2 emissions and the share of coal in the energy mix for global cities

Source: Environment Europe™ Sustainable Cities Database, 2018

Janeiro, Bogota, Quito, Madrid, Adelaide, Copenhagen and Rome have relatively low shares of coal in the energy mix and lower levels of CO_2 emissions per capita.

A significant correlation between CO_2 emissions and the share of trips made by walking, cycling and public transportation has been confirmed (Figure 3.3), which enriches our understanding of these practices and the urban planning tools that encourage them for greenhouse gas emissions.

Such cities as Stockholm, Mumbai, Bogota, Delhi, Mexico City, Paris, Amsterdam, Seoul, Barcelona, Sao Paulo, Berlin, Singapore and Moscow have a significant percentage of trips made by walking, cycling and using public transportation and are associated with lower per capita CO_2 emissions. On the other hand, such cities as Sydney, Shenzhen, Almaty, Los Angeles, Miami, Kuala Lumpur, Boston,

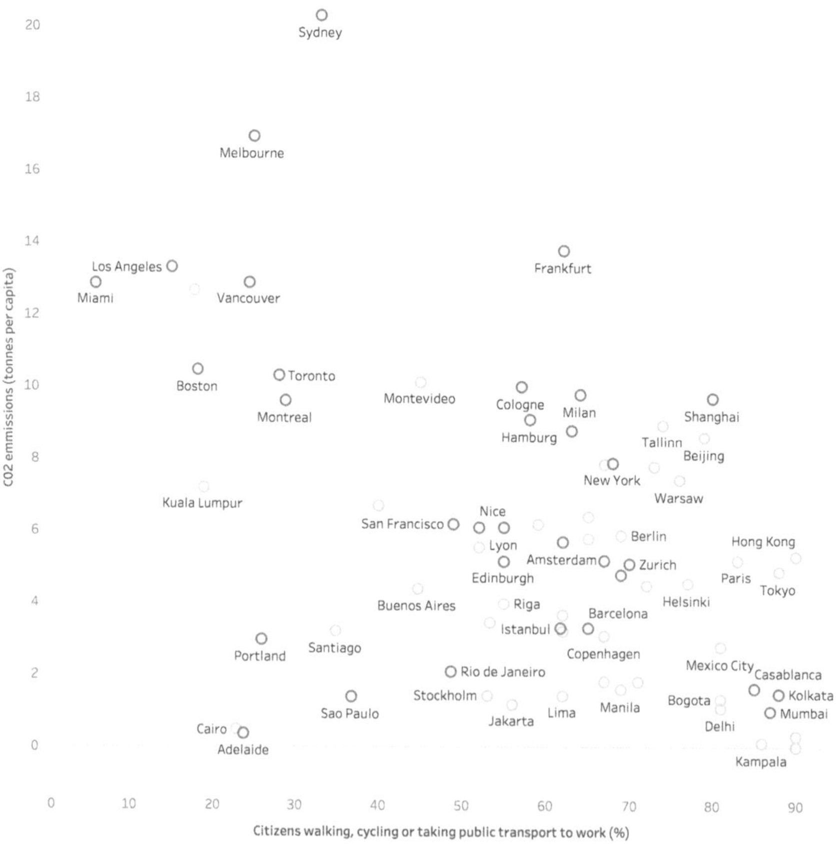

Figure 3.3 Correlation between CO_2 emissions and the share of trips made by walking, cycling and taking public transportation

Source: Environment Europe™ Sustainable Cities Database, 2018

Vancouver and Toronto rely on private cars in a much more pronounced way, and therefore have significantly higher CO_2 emissions per capita.

The role of renewable energy in reducing CO_2 emissions in global cities has been confirmed at a very high level of statistical significance (Figure 3.4). This clearly reinstates the tendency in such cities like Sao Paulo, Bogota, Montreal, Stockholm, Rio de Janeiro, Zurich and Copenhagen that are largely powered by hydro energy to have lower per capita CO_2 emissions. At the same time, cities like Sydney, Atlanta, Almaty, Frankfurt, Miami, St Petersburg, Shanghai, Boston, Los Angeles, Vancouver and Shenzhen, which feature lower levels of renewables in the energy mix, exhibit higher per capita CO_2 emissions.

Based on the data covering 71 cities contained in the Environment Europe Sustainable Cities Database, we were able to generate a regression that captured

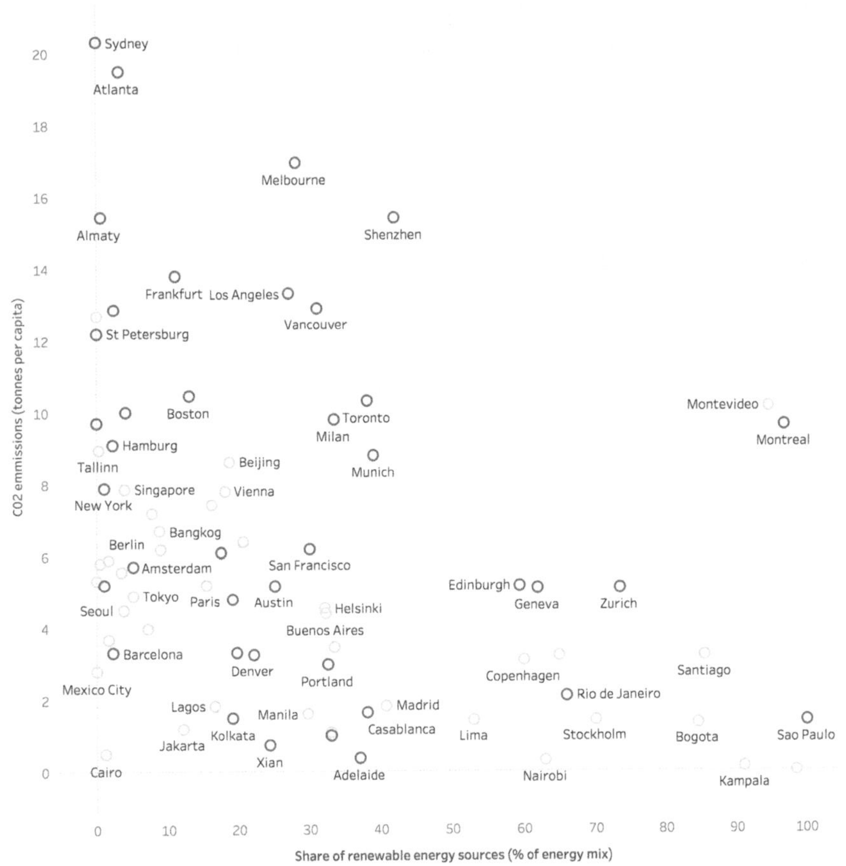

Figure 3.4 Correlation between CO_2 emissions and the share of the renewable energy for global cities

Source: Environment Europe™ Sustainable Cities Database, 2018

80% of the variation in urban CO_2 emissions across the whole world (Table 3.1). Urban CO_2 emissions tend to decrease with the increasing daily mean temperatures in the city. On average, higher temperatures result in reduced need for heating and associated CO_2 emissions. With our current models, unfortunately, we cannot take into the account increased electricity consumption due to air conditioning. One finding that stands out is that cities with OECD capital status tend to exhibit significantly lower CO_2 emissions, possibly as a result of higher technological development in public transportation systems, electric cars and pedestrianization as a new trend in urban planning and design. As the correlations presented here show, a large share of renewables in the energy mix tends to reduce urban CO_2 emissions according to our results. On the other hand, a higher share of coal in the energy mix tends to increase urban CO_2 emissions. The behavioural variable, representing the share of trips made by walking, cycling and using public transportation, reduces urban CO_2 emissions (Table 3.1). Paradoxically, higher recycling rates, everything else being equal, tend to increase CO_2 emissions as additional amounts of energy are needed for complex recycling. The CO_2 tax, reflecting the existing structure of incentives globally, helps to reduce CO_2 emissions, which is statistically significant at 5%.

One of the more interesting variables in the model is the OECD capital status. It is clear from the equation coefficients that capital status implies an average reduction of over 2 tonnes of CO_2 per person per year. What could be a potential explanation of this phenomenon? One hypothesis could be the OECD policy leadership in the global pedestrianization movement. According to Parajuli and Pojani (2017), Copenhagen has been a true pioneer, entirely pedestrianizing Strøget, the central shopping street, in 1962. In Sweden, Stockholm, Lund and Gothenburg managed to pedestrianize their centres by 1961 (Parajuli and Pojani, 2017). The Organization for Economic Cooperation and Development (OECD) introduced new policies aimed at promoting new pedestrian malls in 1972.

Table 3.1 CO_2 regression coefficients in the linear urban regression

Variable	*Coefficient*	*Std. error*	*t-value*	*t-prob*	*Part. R^2*
Constant	15.2640	1.023	14.9	0.0000	0.7794
Daily mean temperature	−0.234784	0.04427	−5.30	0.0000	0.3087
OECD capital status	−2.29855	0.6474	−3.55	0.0007	0.1667
Share of renewables in the energy mix	−0.0376761	0.01115	−3.38	0.0013	0.1534
Share of coal in the energy mix	0.0486420	0.009920	4.90	0.0000	0.2762
Share of trips made by walking, cycling and public transportation	−0.113082	0.01036	−10.9	0.0000	0.6543
Recycling rate	0.0692216	0.01286	5.38	0.0000	0.3150
CO_2 tax	−0.0306765	0.01428	−2.15	0.0355	0.0683

Source: Environment Europe Sustainable Cities Database, 2018

Note: 71 observations; R^2 = 0.805394

According to Feriel (2016), in 1969, the Dutch politician Emile van Lennep became Secretary-General of the OECD and started a new era of increased interest in environmental and urban matters. Attracting attention to "Problems of modern society: economic growth, environment and welfare" as early as 1969, the OECD established a new Urban Environment group within its Environment Committee, which coincided with launching environmental ministries in the UK (1970) and France (1971). The research topics the Urban Environment group has been focusing on providing a good illustration of the direction of thinking at the OECD: "Vehicle-free areas in cities" (1970–1972), "Policy instruments for influencing the form and structure of urban development" (1972–1974), "Urban noise abatement" (1973–1975), "Low-cost improvements in the outdoor urban environment" (1973–1975), "Management of publicly-owned land" (1975–1977) and "Traffic policies for improvement of urban environment" (1975–1976) (Feriel, 2016). The active historical role of the OECD in promoting pedestrianization in urban areas gives us confidence to conclude that it might have something to do with the significant reduction in per capita CO_2 emissions in OECD capitals.

All of these developments are important in the area of capital cities because they facilitate the kind of face-to-face meetings necessary to make a capital city hub work at its highest capacity. The more personal connections that take place in the city, the more likely the city is to develop innovative sustainability policies. This section has demonstrated that a variety of factors make a city sustainable, among them capital city status. In the next section, we examine to what extent the capital status matters.

Sustainability assessment: linear aggregation

As a first step in the multidimensional sustainability assessment, we have carried out a linear aggregation with different weights representing varying policy priorities. We had sufficient data on the respective sustainability indicators for 57 cities. This set includes the most important global cities representing Europe, Africa, Asia, North and South America, and Oceania. The cities in question cover 6.7% of the European population, 3.2% of the population of Asia, 5.4% of the North American population, 10.5% of the South American population, 26.1% of the population of Oceania and 0.7% of the African population, which indicates some disbalance, which we are planning to address in the future by increasing the share of Asian and African cities in the database. The cities in the Environment Europe database include both A++ cities, London and New York, most A+ cities, including Singapore, Shanghai, Tokyo, Hong Kong, Beijing, Paris, and most significant A−, B+, B and B− cities from different regions.

The linear aggregation assumed perfect substitutability among sustainability criteria. Several different policy priorities were applied, placing the emphasis on economic, social, environmental or smart dimensions through varying weights. The results of the assessment applied to 57 global cities are displayed in Table 3.2.

Table 3.2 Multidimensional Sustainability Assessment of Global Cities under environmental, economic, social and smart priorities, linear aggregation

	Equal priorities		*Environmental priorities*		*Economic priorities*		*Social priorities*		*Smart priorities*	
	City	*Score*	*City*	*Score*	*City*	*Score*	*City*	*Score*	*City*	*Score*
1	San Francisco	0.73	San Francisco	0.75	San Francisco	0.71	Stockholm	0.80	Stockholm	0.73
2	Stockholm	0.72	Stockholm	0.73	Tokyo	0.70	Copenhagen	0.74	San Francisco	0.66
3	Seoul	0.68	Seoul	0.69	Seoul	0.70	Zurich	0.73	Paris	0.62
4	Tokyo	0.67	Copenhagen	0.68	Stockholm	0.68	Munich	0.72	Tokyo	0.61
5	Copenhagen	0.66	Zurich	0.67	Washington	0.63	Seoul	0.71	Boston	0.60
6	Zurich	0.65	Tokyo	0.66	Zurich	0.62	San Francisco	0.71	Frankfurt	0.59
7	Munich	0.65	Munich	0.66	Paris	0.62	Madrid	0.71	Copenhagen	0.59
8	Frankfurt	0.62	Montreal	0.64	Munich	0.62	Tokyo	0.70	Seoul	0.58
9	Madrid	0.62	Madrid	0.64	New York	0.62	Vancouver	0.69	Taipei	0.57
10	Paris	0.61	Frankfurt	0.62	Boston	0.61	Toronto	0.68	Munich	0.57
11	Montreal	0.61	Amsterdam	0.62	Copenhagen	0.61	Amsterdam	0.68	Washington	0.55
12	Vancouver	0.60	Toronto	0.62	Taipei	0.60	Barcelona	0.68	Barcelona	0.54
13	Singapore	0.60	Singapore	0.61	Frankfurt	0.60	Frankfurt	0.67	Singapore	0.53
14	Toronto	0.60	Vancouver	0.61	Singapore	0.60	Edinburgh	0.67	Zurich	0.53
15	Amsterdam	0.60	Paris	0.61	Beijing	0.60	Vienna	0.66	Madrid	0.53
16	Boston	0.60	Adelaide	0.60	Austin	0.59	Montreal	0.66	Vancouver	0.51
17	Taipei	0.59	Edinburgh	0.60	Vancouver	0.58	Taipei	0.66	Shenzhen	0.50
18	Vienna	0.58	Vienna	0.60	Portland	0.58	Berlin	0.63	Amsterdam	0.50
19	Barcelona	0.58	Milan	0.60	Toronto	0.58	Warsaw	0.63	Austin	0.48
20	Milan	0.58	Barcelona	0.59	Denver	0.57	Sydney	0.62	Vienna	0.48
21	Washington	0.58	Taipei	0.59	Madrid	0.56	London	0.62	New York	0.48
22	Sydney	0.57	Sydney	0.59	Montreal	0.56	Milan	0.62	Milan	0.47
23	Edinburgh	0.57	Berlin	0.59	Milan	0.56	Paris	0.61	Berlin	0.46
24	Adelaide	0.57	Boston	0.59	Amsterdam	0.55	Shanghai	0.60	Sydney	0.46
25	New York	0.57	London	0.58	Vienna	0.55	Boston	0.60	Hong Kong	0.46
26	Berlin	0.57	Portland	0.58	Atlanta	0.55	Adelaide	0.59	Portland	0.45
27	Portland	0.57	Rome	0.57	London	0.55	Austin	0.59	Montreal	0.45

28	London	0.56	Warsaw	0.57	Sydney	0.55	Rome	0.59	Toronto	0.45
29	Austin	0.56	Washington	0.57	Shenzhen	0.55	Washington	0.58	Atlanta	0.45
30	Warsaw	0.55	New York	0.57	Los Angeles	0.54	Portland	0.57	London	0.43
31	Beijing	0.55	Austin	0.55	Barcelona	0.54	Shenzhen	0.57	Los Angeles	0.41
32	Shenzhen	0.55	Shenzhen	0.55	Melbourne	0.54	Singapore	0.57	Beijing	0.40
33	Rome	0.54	Beijing	0.54	Adelaide	0.53	New York	0.56	Denver	0.40
34	Denver	0.53	Denver	0.53	Edinburgh	0.53	Melbourne	0.56	Warsaw	0.39
35	Los Angeles	0.52	Los Angeles	0.53	Berlin	0.53	Denver	0.55	Edinburgh	0.39
36	Hong Kong	0.51	Sao Paulo	0.53	Warsaw	0.52	Beijing	0.53	Shanghai	0.38
37	Shanghai	0.50	Bogota	0.53	Hong Kong	0.51	Atlanta	0.53	Rome	0.38
38	Atlanta	0.50	Hong Kong	0.52	Rome	0.51	Los Angeles	0.52	Miami	0.38
39	Melbourne	0.49	Quito	0.50	Shanghai	0.50	Moscow	0.50	Adelaide	0.37
40	Sao Paulo	0.49	Shanghai	0.50	Miami	0.49	St Petersburg	0.50	Melbourne	0.37
41	Bogota	0.48	Lima	0.49	Sao Paulo	0.45	Hong Kong	0.49	Moscow	0.35
42	Quito	0.46	Melbourne	0.49	Moscow	0.44	Almaty	0.46	St Petersburg	0.32
43	Lima	0.46	Kampala	0.48	Lima	0.44	Miami	0.45	Sao Paulo	0.32
44	Moscow	0.46	Moscow	0.48	Santiago	0.43	Sao Paulo	0.44	Mexico City	0.30
45	Miami	0.44	Atlanta	0.48	Mexico City	0.43	Lima	0.44	Santiago	0.30
46	Mexico City	0.43	St Petersburg	0.46	Bogota	0.43	Santiago	0.44	Lima	0.28
47	St Petersburg	0.43	Rio de Janeiro	0.45	Quito	0.42	Istanbul	0.42	Bogota	0.28
48	Kampala	0.43	Mexico City	0.45	Rio de Janeiro	0.40	Mexico City	0.41	Istanbul	0.27
49	Santiago	0.43	Istanbul	0.45	Almaty	0.39	Bogota	0.41	Almaty	0.27
50	Rio de Janeiro	0.42	Santiago	0.44	Istanbul	0.38	Quito	0.40	Rio de Janeiro	0.27
51	Istanbul	0.42	Miami	0.43	St Petersburg	0.38	Buenos Aires	0.38	Quito	0.27
52	Mumbai	0.39	Mumbai	0.42	Mumbai	0.37	Kampala	0.38	Buenos Aires	0.26
53	Buenos Aires	0.39	Nairobi	0.42	Buenos Aires	0.37	Rio de Janeiro	0.38	Kampala	0.24
54	Almaty	0.38	Buenos Aires	0.41	Kampala	0.36	Delhi	0.32	Mumbai	0.23
55	Delhi	0.38	Delhi	0.40	Delhi	0.36	Mumbai	0.31	Delhi	0.22
56	Nairobi	0.34	Almaty	0.38	Johannesburg	0.28	Nairobi	0.23	Nairobi	0.18
57	Johannesburg	0.29	Johannesburg	0.32	Nairobi	0.23	Johannesburg	0.17	Johannesburg	0.16

The results clearly show that under environmental priorities the top five cities have been San Francisco, Stockholm, Seoul, Copenhagen and Zurich; under smart policy priorities the leading cities are Stockholm, San Francisco, Paris, Tokyo and Boston. San Francisco leads under environmental and economic priorities, and Stockholm under social and smart priorities. Seoul takes 3rd place globally under environmental priorities, 3rd place under economic, 5th under social and 8th under smart, which is an extremely strong performance overall. Copenhagen occupies the 4th place in the world under environmental priorities, 11th position under economic, 2nd under social and 7th under smart policy priorities. London can be found at 25th place under environmental priorities, 27th under economic, 21st under social and 30th under smart. Washington, DC occupies 29th place under environmental, 5th under economic, 29th under social, and 11th place under smart priorities.

The worst performing cities in our database are Johannesburg, Almaty, Delhi, Buenos Aires and Nairobi under environmental priorities; Nairobi, Buenos Aires, Johannesburg, Delhi and Kampala under economic priorities; Johannesburg, Nairobi, Mumbai, Delhi and Rio de Janeiro under social priorities; and Johannesburg, Nairobi, Delhi and Kampala under smart priorities.

Again, as we saw in the earlier analysis, capital cities are among the top performers, but other global cities also perform well and not all capital cities reach the top of the lists.

Most sustainable global cities

In the following sections, we explore the top three most sustainable and smart cities globally to explain how they achieved their remarkable success. Among the most successful cities are San Francisco, the U.S. high-tech hub across the continent from the capital in Washington, DC, as well as two national capitals: Stockholm and Seoul.

San Francisco

San Francisco, which is part of an OECD country but not a capital, leads our ranking in Economic and Environmental Dimensions worldwide, which corresponds to the World Economic Forum ranking. The Strategic Plan of San Francisco for 2016–2020 defined the city mission as "to provide solutions that advance climate protection and enhance quality of life for all San Franciscans (SF Environment, no date)." The Strategic Plan has five goals: 1) promoting healthy communities and ecosystems; 2) leading on climate action; 3) strengthening community resilience; 4) eliminating waste; 5) amplifying community action.

In particular, Goal 2 uses an active target to reduce greenhouse gas emissions by 40% by 2025 and has the following subgoals: maximize energy efficiency in existing buildings; reduce dependency on single occupancy vehicles by improving access to sustainable and affordable modes of transportation; commit to ambitious carbon reduction targets across city agencies; continue to share San Francisco's practices and lessons to show the world what is possible; decarbonize the energy used for heating and cooling buildings; accelerate the shift to 100% renewable grid electricity by 2030 and maximize local on-site generation of renewable electricity through policy development and investment; and decarbonize the transport sector by facilitating deployment of electric and zero-emissions vehicles.

San Francisco is one of the world leaders on recycling (80%), generating very small amounts of municipal solid waste per person (195.4 kg per year). Nearly half (49%) of the trips made by citizens are carried out by walking, cycling or using public transportation. San Francisco generates 6.2 tonnes of CO_2 per person per year and produces 30% of its energy through renewable sources. In the field of air quality, San Francisco exhibits low levels of PM_{10} pollution at 15.77 $\mu g/m^3$, which is within the World Health Organization limit of 20 $\mu g/m^3$. It has a reasonably extensive system of public transportation.

Economically, San Francisco is one the most vibrant places in the world. With high per capita income of US$88,518 at PPP in 2010 prices, inflation is low at 3.8%, unemployment at 4.4%, a third the level of nearby Los Angeles. San Francisco is a world innovation hub with 3.24 patents registered per 1,000 inhabitants, which is higher than Boston. Income differentiation in San Francisco is high, illustrated by a Gini index of income inequality of 0.51. Such relatively high income inequality could limit San Francisco's performance in the social dimension.

Stockholm

Stockholm, an OECD capital, won the prestigious European Green Capital prize in 2010. The city adopted an Environmental Program for 2016–2019, based on complementarity between environmental protection and human needs. The six priority areas of this programme include: sustainable energy use, environmentally friendly transport; sustainable land and water use, resource efficient recycling, a non-toxic Stockholm and a healthy indoor environment. Our research shows that along with taking the environment seriously, Stockholm exhibits extremely strong economic performance. Sweden is consistently ranked high in the World Economic Forum Global Competitiveness Index. Sweden boasts an open economy and outperforms the USA, Japan and Brazil by attracting approximately 4.7% of GDP in foreign direct investment per annum. At the same time, it invests in the range of 3.7% of GDP in research and development, which is considerably higher than the EU average of 1.8%. Sweden and Stockholm managed to decouple economic development from growth in CO_2 emissions as a result of technological modernization in the 1970s with the extensive use of hydropower and nuclear energy, as well as successful application of environmental taxes since 1991 (Shmelev and Speck, 2018). Stockholm aims to be fossil fuel free by 2050, and is actively involved in new programmes on green urban transport.

The Stockholm economy is largely innovations-based, with the number of new patents registered (2.62 per 1,000 inhabitants) higher than that of all other regional European centres, including technological giants like Copenhagen,

Munich and Zurich. Stockholm outperforms Tokyo. At the same time, Stockholm is characterized by very low inflation, in fact deflation at 0.04%. The level of higher education in Stockholm is 58% of all the residents aged 25–64. Stockholm, representing the Nordic governance model, is characterized by high levels of taxation as a percentage of GDP and a reasonably low Gini index of income inequality (0.3).

In the environmental dimension, Stockholm is characterized by low CO_2 emissions at 1.44 tonnes per capita. One of the possible reasons for such low CO_2 emissions in Stockholm could be its active reliance on renewable energy. Stockholm occupies one of the leading positions in Europe on the share of renewables in the energy mix (70%), following Zurich. Stockholm's performance on renewables is considerably better than other European cities – Copenhagen, Edinburgh, Madrid, Rome, Moscow, Vienna, Paris, London, Amsterdam. On the other hand, according to the data on the share of all trips made by walking, cycling and using public transportation, Stockholm is unfortunately not in the lead, following Vienna, Madrid, Moscow, Amsterdam and London at a modest level of 53%.

Another important parameter for "explaining" low CO_2 emission levels is infrastructure, which gives affordance to using public transportation by residents. In this regard, Stockholm is characterized by a highly diversified system of underground networks with 108 underground stations per 100,000 inhabitants. This availability of public transportation is better than most European cities: Madrid, Amsterdam, London, Rome and Berlin, but not Paris. Air quality in Stockholm is at a good European level, with an average annual concentration of PM_{10} at 26 $\mu g/m^3$, which is nevertheless higher than the maximum recommended by the World Health Organization of 20 $\mu g/m^3$. Better air quality is observed in such European cities as Edinburgh, Madrid, Zurich, Amsterdam and Vienna: it is worse in London and Paris. In the field of circular economy, Stockholm generates rather large amounts of municipal solid waste of 597 kg/person per year, 31% of which is recycled. Other European cities practice less resource-intensive lifestyles: Madrid, Amsterdam, Berlin, London, Paris, Vienna. Recycling rates are lower than Stockholm in Madrid, Rome, Paris and Copenhagen, and higher in Vienna, London, Berlin and Amsterdam.

Seoul

In November 2017, the Metropolitan Government of Seoul, an OECD capital, adopted 17 Sustainable Development Goals and 96 targets. Seoul Plan 2030, an urban planning document, covers three central dimensions: environment, society and culture, and the economy, and includes 30 urban development indicators. Among Seoul's strategic priorities are reducing Seoul's reliance on nuclear power, increasing energy efficiency, and expanding female participation in economic activities. Already in 2013, the International Telecommunication Union issued a Smart Cities report devoted to Seoul's achievements. The Seoul's Smart City programme includes a fast optical-wire and wireless network. Seoul began distributing second-hand smart devices to low-income families and established a "u-Seoul net" in 2003, which connected major public buildings, offices and municipalities via fibre-optic cables arranged along Seoul's underground tunnels. The Smart Work Center was established to allow government employees to work closer to home, and 30% of staff were covered by this initiative in 2015. Seoul's open governance model implies a strong system of community mapping, through which citizens can raise concerns about their neighbourhoods and communities. Seoul's smart metering project aims to reduce electricity consumption by 10%, and in 2012 a pilot project covered 1,000 families with smart meters. The Open Data Square covers information on general administrative work, welfare, culture and tourism, city management, environment, safety, education, health, industry, economy and transportation. Smart solutions are used in Seoul to optimize personal travel of citizens, planning routes, choosing green transport solutions and reducing carbon emissions.

Seoul's metropolitan area hosted nearly 50% of the Korean economy in 2013; at the same time, the Seoul metropolitan area provided employment for 50% of the country's population. Seoul's unemployment rate of 2.3% in 2014 was at the level of regional leaders like Beijing and Singapore, but lower than that of Tokyo. Seoul has a significant rate of residents with higher education (40.6%), which is slightly lower than Singapore but higher than regional centres Beijing, Hong Kong, Shanghai and Shenzhen, and is higher than similar levels in Berlin, Vienna and Rome. The reasonably low Gini index of 0.3 underlines the values of equality in Korean society and is considerably lower than that of regional leaders like Singapore, Beijing and Hong Kong. Inflation in Seoul is low at 0.71%, which is comparable only to Copenhagen, not mentioning deflation in Stockholm. According to our model, which uses the Smart and Sustainable Urban Development Indicator Framework, the number of patents registered per 1,000 inhabitants in Seoul is at a very respectable level of 1.4 per year.

In the environmental field, CO_2 emissions per capita measured in Seoul on an annual basis at 4.5 are lower than regional leaders Tokyo, Hong Kong, Singapore, Beijing, Shanghai and Shenzhen. Compared with European cities, Seoul trails Scandinavian cities that traditionally exhibit very high performance, such as Stockholm and Copenhagen, but is performing better than Paris, Amsterdam, Berlin, London, Vienna and Munich. Seoul also outperforms San Francisco, New York, Montreal, Boston, Washington and Los Angeles.

Recycling is definitely one of the main strengths of Seoul, with 63.5% of all collected municipal solid waste being recycled. Green space in Seoul is not particularly abundant at 1.39 m^2 per person, which is lower than Tokyo, Beijing, Shanghai, Barcelona, London, Paris, Stockholm, Berlin, Rome and Copenhagen.

Conclusion

This chapter examined capital cities as a subset of global cities – the centres for economic activity, and the cities that are responsible for a considerable share in global CO_2 emissions and produce substantial volumes of waste. The discussion laid out a variety of indicators that can be used to measure the performance of capital cities.

The application of cross-sectional regression allowed us to produce a robust CO_2 emission model for these cities. The key factors affecting the CO_2 emissions we identified are: the share of coal in the energy mix; share of renewables; share of trips made by walking, cycling and public transportation; mean annual temperature; OECD capital status; recycling rates; and CO_2 taxes. All the factors combined explain 80% of the variation in urban CO_2 emissions worldwide. The multidimensional sustainability assessment identified three clear sustainability leaders: San Francisco, Stockholm and Seoul. The results placed the performance of individual cities within the global context and presented the indicator-based sustainable development performance of individual cities within a coherent framework.

It is clear that a multitude of city types are among the top most sustainable cities in the world: national political capitals Stockholm, Seoul, Copenhagen, Tokyo, Madrid, Singapore, Amsterdam, Paris and Vienna; financial hubs Zurich, Frankfurt and Toronto; innovation hubs San Francisco, Munich and Vancouver; and regional hubs Montreal, Barcelona and Edinburgh – which corresponds to the classification outlined by the European Commission (2007). OECD capital status played a significant role in explaining CO_2 emissions, one of the key urban sustainability indicators.

Overall, the research suggested that beyond the focus on walkability and public transportation, little distinguishes capital cities from other global cities. Capital cities often perform well in sustainability measures, but not always, and other cities play a leadership role, as well. This finding suggests that capital cities can do more to promote sustainability policies and practices. Further development of urban sustainability indicators will allow us to better measure performance of the cities and identify the best practices that can be transferred among cities.

References

Ahvenniemi, H., Huovila, A., Pinto-Seppä, I., Airaksinen, M. (2017). What Are the Differences Between Sustainable and Smart Cities? *Cities* 60, Part A, 234–245.

Bai, X. et al. (2018). Six Research Priorities for Cities and Climate Change, *Nature*, 555, 23–25.

Bithas K. P., Christofakis, M. (2006). Environmentally Sustainable Cities: Critical Review and Operational Conditions, *Sustainable Development*, 14, 177–189.

Dassen, T., Kunseler, E., van Kessenich, L.M. (2013). The Sustainable City: An Analytical-Deliberative Approach to Assess Policy in the Context of Sustainable Urban Development, *Sustainable Development*, 21, 193–205.

EC. (2009). *Sustainable Development Indicators: An Overview of relevant Framework Programme Funded Research and Identification of Further Needs in View of EU and International Activities*, Brussels: European Commission.

European Commission. (2007). *State of European Cities Report, Adding Value to European Urban Audit*, http://ec.europa.eu/regional_policy/sources/docgener/studies/pdf/urban/stateofcities_2007.pdf.

Eurostat. (2016). *Urban Europe: Statistics on Cities, Towns and Suburbs*, http://ec.europa.eu/eurostat/web/products-statistical-books/-/KS-01-16-691.
Feriel, C. (2016). Better (Urban) Policies for Better Lifes – The Role of OECD in Transnational Exchanges of Planning Ideas in the 1970s, in Hein, C. ed. *International Planning History Society Proceedings, 17th IPHS Conference, History-Urbanism-Resilience, TU Delft 17–21 July 2016*, V.06, TU Delft Open.
García-Fuentes, M.Á., Quijano, A., De Torre, C., García, R., Compere, P., Degard, C., Tomé, I. (2017). European Cities Characterization as Basis Towards the Replication of a Smart and Sustainable Urban Regeneration Model, *Energy Procedia*, 111, 836–845.
Girardet H. (1993). *The Gaia Atlas of Cities: New Directions for Sustainable Urban Living*, London: Gaia Books Ltd.
Girardet, H. (2004). *Cities People Planet: Liveable Cities for a Sustainable World*, New York: John Wiley & Sons.
Girardet, H. (2014). *Creating Regenerative Cities*, London: Routledge.
Girardi, P., Temporelli, A. (2017). Smartainability: A Methodology for Assessing the Sustainability of the Smart City, *Energy Procedia*, 111, 810–816.
Hall, P. (2014). *Good Cities, Better Lives: How Europe Discovered the Lost Art of Urbanism*, London: Routledge.
Hall, P., Pfeiffer, U. (2000). *Urban Future 21: A Global Agenda for Twenty-First Century Cities*, London: Routledge.
Hara, M., Nagao, T., Hannoe, S., Nakamura, J. (2016). New key Performance Indicators for a Smart Sustainable City, *Sustainability*, 8(3), Article number 206.
ISO, 2014 ISO 37120:2014. (en) *Sustainable Development of Communities – Indicators for City Services and Quality of Life*. Washington, DC: International Organization for Standards.
Kierstead, J., Leach, M. (2008). Bridging the Gaps Between Theory and Practice: A Service Niche Approach to Urban Sustainability Indicators, *Sustainable Development*, 16, 329–340.
Klopp, J.M., Petretta, D.L. (2017). The Urban Sustainable Development Goal: Indicators, Complexity and the Politics of Measuring Cities, *Cities*, 63, 92–97.
Manitiu, D.N., Pedrini, G. (2016). Urban Smartness and Sustainability in Europe: An Ex Ante Assessment of Environmental, Social and Cultural Domains, *European Planning Studies*, 24(10), 1766–1787.
Martin, N., Rice, J. (2014). Sustainable Development Pathways: Determining Socially Constructed Visions for Cities, *Sustainable Development*, 22, 391–403.
Monfaredzadeh, T., Berardi, U. (2015). Beneath the Smart City: Dichotomy Between Sustainability and Competitiveness, *International Journal of Sustainable Building Technology and Urban Development*, 6(3), 140–156.
Mori, K., Yamashita, T. (2015). Methodological Framework of Sustainability Assessment in City Sustainability Index (CSI): A concept of Constraint and Maximisation Indicators, *Habitat International*, 45, 10–14.
Naess, P. (1995). Central Dimensions in a Sustainable Urban Development, *Sustainable Development*, 3, 120–129.
Parajuli, A., Pojani, D. (2017). Barriers to the Pedestrianization of City Centres: Perspectives from the Global North and the Global South, *Journal of Urban Design*, 23, 142–160.
Pierce, P., Ricciardi, F., Zardini, A. (2017). Smart Cities as Organizational Fields: A Framework for Mapping Sustainability-Enabling Configurations, *Sustainability* (Switzerland), 9 (1506).

Roy, B. (1996). *Multicriteria Methodology for Decision Aiding*, Amsterdam: Kluwer Academic Publishers.

SF Environment. (no date). *Strategic Plan 2016–2020*. https://plan.sfenvironment.org/. Accessed January 3, 2019.

Shen, L-Y., Zhou, J. (2014). Examining the Effectiveness of Indicators for Guiding Sustainable Urbanization in China, *Habitat International* 44, 111–120.

Shmelev, S.E. (2010). *Environmentally Extended Input-Output Analysis of the UK Economy: Key Sector Analysis*, Oxford: University of Oxford.

Shmelev, S.E. (2011). Dynamic Sustainability Assessment: The Case of Russia in the Period of Transition (1985–2008), *Ecological Economics*, 70(11), 2039–2049.

Shmelev, S.E. (2017). Multidimensional Sustainability Assessment for Megacities, in Shmelev, S. ed. *Green Economy Reader: Lectures in Ecological Economics and Sustainability*, Berlin: Springer, 205–236.

Shmelev, S.E., Shmeleva, I.A. (2009). Sustainable Cities: Problems of Integrated Interdisciplinary Research. *International Journal of Sustainable Development*, 12(1), 4–23.

Shmelev S.E., Speck, S.U. (2018). Green Fiscal Reform in Sweden: Econometric Assessment of the Carbon and Energy Taxation Scheme, *Renewable & Sustainable Energy Review*, 90, 969–981.

Spangenberg, J.H. (2002a). Environmental Space and the Prism of Sustainability: Frameworks for Indicators Measuring Sustainable Development, *Ecological Indicators*, 2, 295–309.

Spangenberg, J.H. (2002b). Institutional Sustainability Indicators: An Analysis of the Institutions in Agenda 21 and a Draft Set of Indicators for Monitoring Their Effectivity, *Sustainable Development*, 10, 103–115.

Spangenberg, J.H. (2005). Economic Sustainability of the Economy: Concepts and Indicators, *International Journal of Sustainable Development*, 8(1/2).

Spangenberg, J.H. (2017). Hot Air or Comprehensive Progress? A Critical Assessment of the SDGs, *Sustainable Development*, 25, 311–321.

UN. (2007). *Indicators of Sustainable Development: Guidelines and Methodologies*, New York: UN.

UN. (2015a). *Technical Report by the Bureau of the United Nations Statistical Commission (UNSC) on the Process of the Development of an Indicator Framework for the Goals and Targets of the Post-2015 Development Agenda*, Working draft.

UN. (2015b). *Transforming our World: The 2030 Agenda for Sustainable Development*, Resolution adopted by the General Assembly on 25 September, A/RES/70/1.

UNECE. (2013). *Framework and Suggested Indicators to Measure Sustainable Development*, Prepared by the Joint UNECE/Eurostat/OECD Task Force on Measuring Sustainable Development 27 May.

UNECE and ITU. (2016). Rome Declaration Adopted by the participants of the Forum "Shaping Smarter and More Sustainable Cities: Striving for Sustainable Development Goals", on 19 May in Rome, www.itu.int/en/ITU-T/Workshops-and-Seminars/Documents/Forum-on-SSC-UNECE-ITU-18-19-May-2016/Rome-Declaration-19May2016.pdf

UN ECOSOC. (2015). *The UNECE – ITU Smart Sustainable Cities Indicators*, www.unece.org/fileadmin/DAM/hlm/documents/2015/ECE_HBP_2015_4.en.pdf

UN Habitat. (2016). *Urbanisation and Development: Emerging Futures*.

UNEP. (2011). *Towards a Green Economy. Pathways to Sustainable Development and Poverty Eradication*, New York: UN Environment Program.

Valentin, A., Spangenberg, J.H. (2000). A Guide to Community Sustainability Indicators, *Environmental Impact Assessment Review*, 20, 381–392.

Wei, Y., Huang, C., Lam, P.T.I., Yuan, Z. (2015). Sustainable Urban Development: A Review on Urban Carrying Capacity Assessment, *Habitat International*, 46, 64–71.

Wei, Y., Huang, C., Li, J., Xie, L. (2016). An Evaluation Model for Urban Carrying Capacity: A Case Study of China's Mega-Cities, *Habitat International*, 53, 87–96.

Wong, C. (2015). A framework for 'City Prosperity Index': Linking Indicators, Analysis and Policy, *Habitat International*, 45, 3–9.

Yigitcanlar, T., Dur, F., Dizdaroglu, D. (2015). Towards Prosperous Sustainable Cities: A Multiscalar Urban Sustainability Assessment Approach, *Habitat International*, 45, 36–46.

Part II

Case studies

4 Moving the capital to promote sustainability innovations

Analyzing Cairo's future plans

Sahar Attia and Costis Toregas

It would be difficult to publish a book about world capitals and not include Cairo. Despite its diverse problems and the poor quality of life of its citizens, Cairo remains one of the leading capitals in the Middle East. In this chapter, the authors investigate and debate what place Cairo occupies among other capital cities in promoting sustainability, what stakeholders are doing to achieve sustainability, and how moving key capital city functions out of Cairo would affect the old and newly created cities in the region.

In the last two decades, many capital cities embarked on planning efforts that laid out visions for the future, while they were actually built and developed without the concept of future visions. Many of these visions focus on sustainability, and use it as a theme for branding their efforts. Linking the past to a future vision in a sustainable context provides a key opportunity for the best use of resources. All stakeholders, including national and local governments, universities, the private sector, and the local community, agree that achieving sustainability is not an option, but a necessity: the shift has to happen, but requires coherent policies and a balance between top-down and bottom-up approaches, a challenging cultural prospect in the Cairo region. Stakeholders have to work together and be supportive of each other in order to face current and future challenges.

This case study chapter explores the hub function of capital cities using Egypt's capital Cairo as the focus of analysis. Today, Egypt's most important city dominates the rest of the country in terms of the size of its economy, services, population, and the use of resources. However, the city is currently burdened by too many people living in close proximity without the necessary services to support them. Moreover, additional migrants continue to move to the city and build houses in informal developments that put an even greater strain on existing municipal services. In order to improve their effectiveness and sustainability, the Egyptian government is in the process of moving key capital administration functions, financial institutions, innovation centers, and embassies out of Cairo to a new city called the "New Administrative Capital" (NAC). In contrast to the current congested and crowded metropolis, the new city is planned to incorporate smart features with high levels of efficient mobility, while Cairo will remain the country's unique cultural and tourism hub, relying on its main original assets and resources.

This chapter illuminates and gives texture to the main theme of the book by examining whether it is possible to extract essential hub functions from an existing city and transfer them elsewhere. Countries such as Brazil and Kazakhstan (among others) have moved their capitals in the past. The chapter examines how Cairo is inspired by other international examples and what it might teach other capital cities where there is a discussion about moving capital city functions within a context of sustainability. Additionally, the chapter highlights which elements are missing and which parts of the hub are working in Cairo, within its context full of contradictions and diversity. Finally, using a case study example abstracted from a pilot project undertaken in 2013–2015, the chapter examines how the newly created city can relate to the "old" city and the people left behind, showing that the problems of informal development and crowding can be overcome through processes engaging the population. This example provides hope that a win-win situation can be found for the new and old cities.

The chapter consists of four sections. The first introduces the debate about capitals around the world, their key functions as hubs, and how moves in the capital have exposed different sides of the hub framework. The second section discusses the long history of Cairo as a capital and examines how the hub works there. The third section debates the raison d'etre of the new administrative capital and the action of constructing a new city versus trying to make the existing capital work more effectively. In particular, the question of the impact of dividing the government and embassies from the historical/cultural section is one that deserves treatment. The fourth section tackles the role of stakeholders in proposing initiatives to reach sustainability in the current capital, as well as support practical solutions to the overcrowding issue. In this context, the case of the Embaba informal development proves that actions can be carried out to improve the situation in the old city by involving residents as part of the process. Finally, we discuss the potential of future work and the inevitability of hybrid solutions.

Capital cities as hubs, and the case for Cairo

More than 40 countries throughout the world have moved their capital cities over the last 200 years. Although there is no clear overriding reason for such moves, a brief exploration of a few of these moves reveals a small number of common, dominant reasons for this action:

1 Splitting the difference between two powerful, competing urban centers. In Canada and Australia, strong, established urban interests compromised by creating new capital cities and consolidating functions in the boundary between the two. Sydney and Melbourne agreed to create a new, planned city that became Canberra in 1927, while Toronto and Quebec decided to split the distance difference between them and established Ottawa as the capital in 1857.
2 Centralizing the location of the capital to make it more accessible to all parts of the country. Brazil faced overcrowding and a perception of unacceptable

distances for travelers seeking government access in Rio de Janeiro's coastal location and created the new capital Brasilia by planning a development hub in the center of the country in 1960.

3 Expectations of political, social, or economic benefit. The move of Kazakhstan's capital from Almaty to Astana was made in an effort to reduce the chances of separatism in Kazakhstan's north by moving more Kazakhs into the region dominated by Slavs near the Russian border and to stimulate economic growth in a relatively poor part of the country.

Many of the historical capital city moves left large numbers of residents behind and anchored the move by transferring governmental functions to the new sites. These transfers by necessity brought along people impacted by secondary linkages to the service industry, transportation sector, and other key elements of the economy beyond government. The notion of resource hubs and sustainability as a driver for a move is not evident from historical activities that can be documented and observed.

The obvious question to address is whether the move left the prior capital city to wither and disintegrate as a vibrant urban center, or whether both old and new capitals were able to thrive. In the earlier examples, both prior and current capitals are today exhibiting strength, implying that an equilibrium of sorts in the hub system has been reached in both places over time. In the case of Rio, even without the governmental functions, the cultural, business and innovation economies have continued to expand in the prior capital. Brasilia has itself grown, but some say it lacks the cultural vibrancy of Rio. While it is not easy to make detailed assessments, it is safe to assume that moving governmental functions does not necessarily eliminate the creative hub element interactions in the prior capital.

However, a capital move *does* give a government the chance to organize the resources in the new location and ensure a priori the existence of a sustainability balance that can be accomplished with less financial and political effort than in the prior overcrowded and unsustainable capital. Understandings of and support for sustainability has grown over the last few decades, giving government planners a chance to consider embedding sustainability in the built environment, the infrastructure and the choices for growth as they lay out a new city. It is this freedom to create more dynamic and pronounced ways to tie the hub elements in a sustainable manner that leads to high expectations as the Egyptian government considers a move of the administrative and political capital away from the current center of Cairo.

The hub definition provided in the introduction of the book applies across many cultures. However, each nation and capital city will emphasize elements of the hub that relate to its history and reflect socially acceptable norms. In Cairo, the main actors of the hub include:

1 The government (national and local)
2 International funding agencies
3 Non-governmental organizations (NGOs)

4 The media
5 Activists and philanthropists
6 Universities and research institutions
7 Civil society
8 The private sector, including investors and developers

Given the urban densities, the constant demand for resources that is hard to meet, and the pressures of unsustainable growth, the role of civil society is a major element of Cairo's resource pool. Engaging civil society is key to creating conditions of change, especially in the sustainability domain. However, there is constant debate about the focus and ultimate objective, and targets of each group of actors that represent the main challenge for achieving the sustainability of the hub elements in Cairo. These elements include: coordination among all actors, transparency in the decision-making process, capacity building of local government, resource efficiencies, the use of technology in monitoring, and many other elements that ensure the functioning of the hub. Moreover, the impact of the development of the NAC is in the center of the debate: will there be an interchange of hubs between Cairo and the NAC?

Cairo's history viewed in a sustainability framework

Cairo[1] is the host of multilayered functions, urban transformations, and a capital city hub. It can be seen as a compilation of history, policies, conflicts, and contradictions. The wealth of challenges, problems, and issues have transformed this megacity into an interesting laboratory over time and made it an attractive and fruitful place to study urban evolution. Cairo, the capital, the city lights, the high-end shopping arcades, is also Cairo the informal, the social disorder, and the bureaucracy (Attia, 2012). Cairo is a city where growth has been aggressive, showing an unbalanced and unplanned rate of development, both spatially and socially. Nevertheless, it is still the main urban center in Egypt, and has been through its history, accumulating cultural, urban, and architectural wealth. In fact, the morphological layers and functions date back since the foundation of Al Fustat (AD 647). The name of Cairo was given by the Fatimids (10th century). Mamluks (14th century) and Ottomans (17th century) ruled Cairo, as well, creating and leaving behind a rich Islamic architectural heritage.

It was only under Mohamed Ali Pasha and his successors in the mid-19th century that Cairo entered into a process of modernization and economic growth, and began to reassert itself politically, based on a large group of European technicians and entrepreneurs. Entrepreneurial minorities such as Greeks, Copts, Levantines, and Armenians figured importantly in economic life and gave impetus to significant growth in economic activity (Sims, 2003). On the other hand, the French and British colonization periods had a great impact on Cairo's urban morphology, architecture, administrative processes, and culture and educational systems. With the 1952 revolution, Cairo witnessed new expansions, and the rise of a new type of urban and architectural features.

Passing through multiple political and economic crises, wars, the death of Anwar El Sadat, the 2011 uprising, and the 2013 June revolution, along with the expansion of informal areas, Cairo remained a strong capital. However, this strength was not successfully reflected in all the components of its capital city hub. Some assets were "winners" and others "losers," creating an imbalance which remains today. And despite all the challenges, the megacity witnessed continuously large-scale developments, national projects, and infrastructure works.

In the late 1970s, Egypt adopted a policy of establishing new cities to accomplish national aims. Three generations of new cities evolved from this decision. New cities were conceived as the best solution to avoid both encroaching on agricultural land and expanding informal development, which challenged infrastructure management, long-term planning and efforts to finance growth. The 1980s witnessed the foundation of the "New Communities" Authority, which initiated its ambitious new cities program for Cairo, starting with cities named for the Tenth of Ramadan and the Sixth of October, followed by Al-Sadat, Borg El Arab, New Salheya, New Damietta and 15th of May. Later, other cities joined this incessant march of expansion.

Although even as the first generation of new cities was built, growth continued to push informal urban forms in an aggressive manner. Unplanned areas were expanding on the fringe areas of Cairo, providing affordable housing but without planned service delivery, only informal transportation, and thereby creating unsustainable communities.

Cairo today operates as a city-state. It is composed of urban and rural areas distributed across three governorates (Cairo, parts of Giza, and parts of Qaliubeya) that form the Greater Cairo Region (GCR). Cairo is the mirror of Egypt, gathering all types of problems, issues, and challenges. The capital city's growth has been rapid in the extreme and still continues unabated, but the maximum population capacity finally may have been reached. According to the 2017 census, the population of the Greater Cairo Region is 23.7 million inhabitants, which is a *quarter* of Egypt's current population of 94.79 million inhabitants (CAPMAS, 2017a, 2017b).

Cairo is by far the "main city" in Egypt, and the most attractive to migrants. Alexandria, which is the second largest city, is only 53 percent the size of Cairo. The population of Cairo is distinguished by its youth. Over a third of the population is under 15 years of age. The sex ratio is similar to that of the country (48% female). The total population statistic for Greater Cairo can differ considerably, depending on the geographical definitions applied (Sims, 2003), and this variability is clearly shown in Table 4.1 (CAPMAS, 2017a and 2017b).

Cairo supports major service provision functions at extremely high population concentration rates when compared to the rest of the country. According to the General Organization for Physical Planning (GOPP), 90 percent of Egypt's cultural facilities, 22.5 percent of GDP, 49 percent of industrial jobs, 40 percent of private sector jobs, 43 percent of public sector employees, over 40 percent of investments, and over 50 percent of Egypt's health and educational facilities are present in Cairo today (GOPP, 2012). Accordingly, it is clear that Cairo hosts

Table 4.1 Greater Cairo Region population, in millions

Census years	*Greater Cairo Region population (in millions)*				*% of Egypt's total population*
	in Cairo Governorate	*in Giza Governorate**	*in Qaliubia Governorate**	*Total GC Region**	
1947	2.062	0.668	0.281	3.013	12.50%
1960	3.358	1.118	0.434	4.910	15.70%
1966	4.232	1.420	0.560	6.211	17.40%
1976	5.074	2.137	0.879	8.090	18.50%
1986	6.069	3.332	1.460	10.860	18.20%
1999	6.789	4.273	2.081	13.144	17.30%
2006	8.402	5.694	4.249	18.345	25.30%
2017	9.539	8.632	5.627	23.798	25.10%

Note: *Urban and rural areas

Source: Sims, 2003 and Census of Egypt, 2017, compiled by the authors

major components of Egyptian life, whether related to service, economy, real estate, administration, and stakeholder components. The three generations of new cities described earlier did not succeed in decentralizing functions and reducing the centrality of the capital. Especially resistant to this decentralization effort was the ongoing informal development, which the new cities sought to stem. The government plan to offer new housing opportunities and encourage demographic dispersion, creating new grounds for formal and planned extensions, was not achieved at the expected rates. Therefore, it was essential to rethink the role of the capital within its boundaries, raising the question of whether the new capital would be an autonomous entity, or simply a new extension of the existing agglomeration.

Institutionally, it is evident that there are strong centralization forces and top-down decision-making processes, creating disparities within the different strata of Cairene society. Moreover, the gap between peoples' aspirations and the decision-making process, lead to disparity of priorities, and raise conflicting drivers of change: concerns about poverty, and endless solutions to limit the informal and unplanned urban extensions.

According to Sims, there are seven large unplanned areas with a population of more than 500,000 inhabitants in each area Figure 4.1 (ISDF, 2017) shows the location of large informal areas. These areas have been there for decades, functioning, surviving, vibrant, but lacking an acceptable quality of life (Werthmann & Bridger, 2015).

Hall and Pfeiffer distinguish among three types of cities around the world:

1 Cities coping with informal hyper growth
2 Cities coping with dynamic growth
3 Weakening mature cities coping with aging

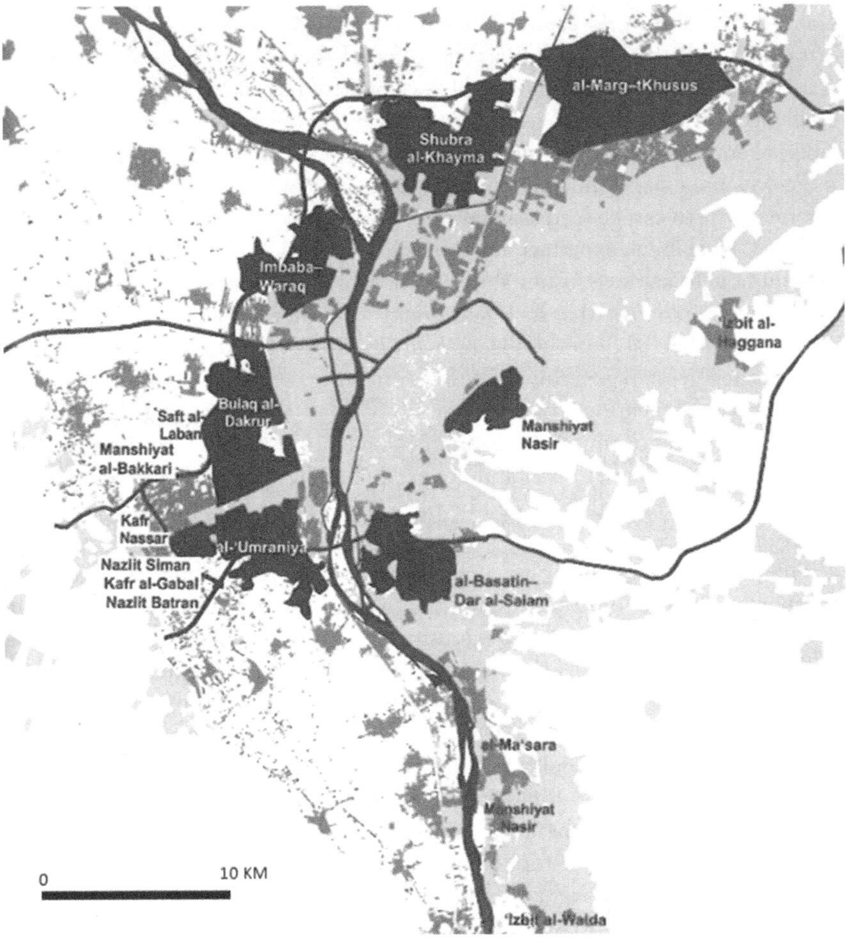

Figure 4.1 Unplanned areas in the Greater Cairo Region

Source: Informal Settlements Development Fund, 2017

Cairo falls within the first type, characterized by rapid population growth, an economy dependent on the informal sector, widespread informal housing and poverty, public health problems, and issues in governance, while the formal sector is struggling to compete (Hall & Pfeiffer, 2000).

Currently, there is strong interest in upgrading all informal and unplanned areas, leading to a better understating of the role residents can play in helping informal areas become more sustainable, thereby lessening the pressure to act in unsustainable ways. All stakeholders are involved, creating sub-hubs for the informal sector. The international funding agencies are also involved, working to facilitate solutions for sanitation, water supply, and waste treatment facilities.

The push for a new administrative Cairo

Cairo, like many world capital cities, is overcrowded as people continue to be attracted to all the major hub activities found there. In order to accommodate this in-migration, significant informal development has taken place, creating more challenges for the planners as well as the operational managers of city services. This pressure has created a political initiative to move the capital and create a new, more organized city in new terrain. The sheer population size and uncontrollable informal growth can be seen as major reasons for this demand for a new capital.

The UN Habitat expert group meeting held in Quito in 2016 discussed in general how to achieve sustainability in new cities in developing countries, and the participants agreed that such new urban areas offer special possibilities as they can function like laboratories for innovation (Provoost, 2016). One of the main reasons the Egyptian government cited for moving the administrative capital was the necessity of creating a new generation of smart and sustainable cities, focusing on de-concentrating the existing agglomerations and controlling informal expansions in the agricultural and desert lands. The increasing demand for new urban areas is a need that should be met, either for the capital with its hub or other important cities.

Building a new administrative capital is not only responding to the demand for increased space in Egypt's capital, but also to the various problems in Cairo as revealed by the Ministry of Housing, Utilities, and Urban Communities (MHUUC). According to the New Urban Communities Authority (NUCA), the new capital is an essential action designed to reduce the problems in the current capital. Initially, the GOPP confirmed that all proposed regeneration and upgrading projects would support the revitalization of Cairo, while creating new engines for economic and social development. In March 2014, the Egyptian government announced the development of a new administrative and financial capital as a solution to restore Cairo's charm as a world class cultural capital. This announcement brought to the forefront both the difficulty of dealing with existing stresses in Cairo, as well as identifying specific functions to be moved to the new capital. Accordingly, a presidential decree was issued to create a new administrative capital to transfer all ministries, embassies, and the main investment and financial institutions, as a solution to resolve issues resulting from the hyper-density and centralization in Cairo.

The future of the Greater Cairo Region has been discussed at many levels of government. Since 2008, several schemes and strategies have been proposed to solve the urgent problems which emerged from rapid growth and limited formal service delivery structures, in an effort to achieve environmental, social, and economic sustainability. Certainly, the residents and government all would like to return Cairo to its past glory again, so it could be one of the most beautiful cities in the world. The challenges that make progress difficult can be summarized along four dominant dimensions:

- Problems resulting from population and economic expansion
- The informal growth of housing and unplanned services

- Meeting environmental obligations and achieving sustainability
- Developing a successful governance system

These challenges figure prominently in the Ministry of Planning, Monitoring and Administrative Reform's statement "Egypt's Vision 2030," which proclaimed that:

> The new Egypt will possess a competitive, balanced and diversified economy, dependent on innovation and knowledge, based on justice, social integrity and participation, characterized by a balanced and diversified ecological collaboration system, investing in the ingenuity of place and humans to achieve sustainable development and to improve Egyptians' life quality (Egypt Vision 2030, No date).

The capital city vision relies on four main pillars: Social Justice, Knowledge and Innovation, Economic Development, and Environmental Concerns. Accordingly, eight moves were suggested to achieve the vision for the capital:

1. Develop social equity by improving living conditions for all
2. Enhance the region's public transportation infrastructure
3. Revitalize the inner city
4. Strengthen the New Urban Communities as vibrant and diverse hubs
5. Create conditions for tourism to flourish while conserving the historic and archeological sites
6. Provide a competitive environment for the knowledge-based economy
7. Improve environmental conditions and achieve sustainability
8. Set-up efficient governance to ensure successful management for the development projects (GOPP, 2012)

The most challenging target is related to sustainability. Cairo's high level of pollution has a dramatic impact on the health of the inhabitants and their quality of life, and also on the economic performance of the region and its ability to attract international institutions and investments. Growing concerns about climate change and the depletion of natural resources – not only in Egypt, but globally – adds to the pressure to act. A reduction in air pollution, attention to water quality, the need for green spaces as well as the efficient management of natural resources (water in particular), will be considered of highest priority in all decisions being made regarding the capital. This provides hope that the pilot discussed below in this chapter regarding water in the informal areas will be given additional attention by the authorities. Sustainability will also support further economic development and promote inclusion (UN Habitat, 2011).

The vision for the Greater Cairo Region sought to exploit its status as the center of governance, institutes of higher learning, and national and international investments, while simultaneously avoiding any moves that would attract more population to its congested geography. Other development processes can be directed toward other parts of the country, based on their current resources and capabilities, by taking into consideration the necessity to provide a balance between all regions in the republic. This policy will maintain a prominent role for the Cairo region, but at the same time create economic centers in other parts of the country with job opportunities and improved living standards for the people. To deliver such a vision, two major strategies are conceived:

- Identify and create a network of investment projects (cultural, administrative, tourism, religious, social) aiming to support and finance the economic development of the strategic vision of the Greater Cairo Region.
- Achieve the state's ambitious project plan and its highest priority, which involves transferring ministries and public buildings from downtown Cairo to the center of the new government area to be located on the edge of the region; also, to link it to a network of roads and transportation links that will make it easier for citizens to move to and from the government buildings from other parts of the city.

In 2010, the General Organization for Physical Planning issued "Egypt 2052" as a national vision (Ministry of Housing, Utilities and Urban Communities, 2014). Development of the fourth generation of new cities was presented as a vital need to conquer the desert and increase livable areas. However, development of the Egyptian desert has faced many challenges that have not been taken into account and addressed. Experts point out that gaps in the previous generations of new towns in Egypt resulted primarily from the absence of understanding the holistic nature of the desert environment and the nature of its interconnections with social, economic, and political systems (Ali, 2003). The lack of a systematic discussion involving stakeholders and experts in ecosystems may have led to this weakness in planning platforms and solutions.

The establishment of new towns in the desert is one of the highest priorities for Egyptian development. In order to enable such construction, planners should first study the historical evolution of new towns that were previously built and track their development in the Egyptian desert over time. It is also essential to monitor the most significant problems associated with planning processes, especially in the Sixth of October new town, one of the largest, as a case study in order to guide potential proposals that may support and promote the achievement of urban development goals in new towns (Hegazy & Moustafa, 2013). It is hoped that policy makers trying to address this important question will tap the expertise and resources of universities, their students and the informal community of stakeholders to help reach decisions that optimize the way forward.

The new administrative capital of Cairo is to be located on approximately 184,000 km^2 as an extension of the city on the east (a 90-minute drive from the center). Currently, some activities are underway in the new capital; they include large-scale projects by both government and the private sector focused on social housing, industrial parks, high-end residential compounds, a robust road network, services, power plants, universities, a sport city, and other functions and assets. This first phase is intended to extend over 184 km^2. The expected population for the city once fully developed is seven million inhabitants. Figure 4.2 presents the general location and transport linkages of the NAC.

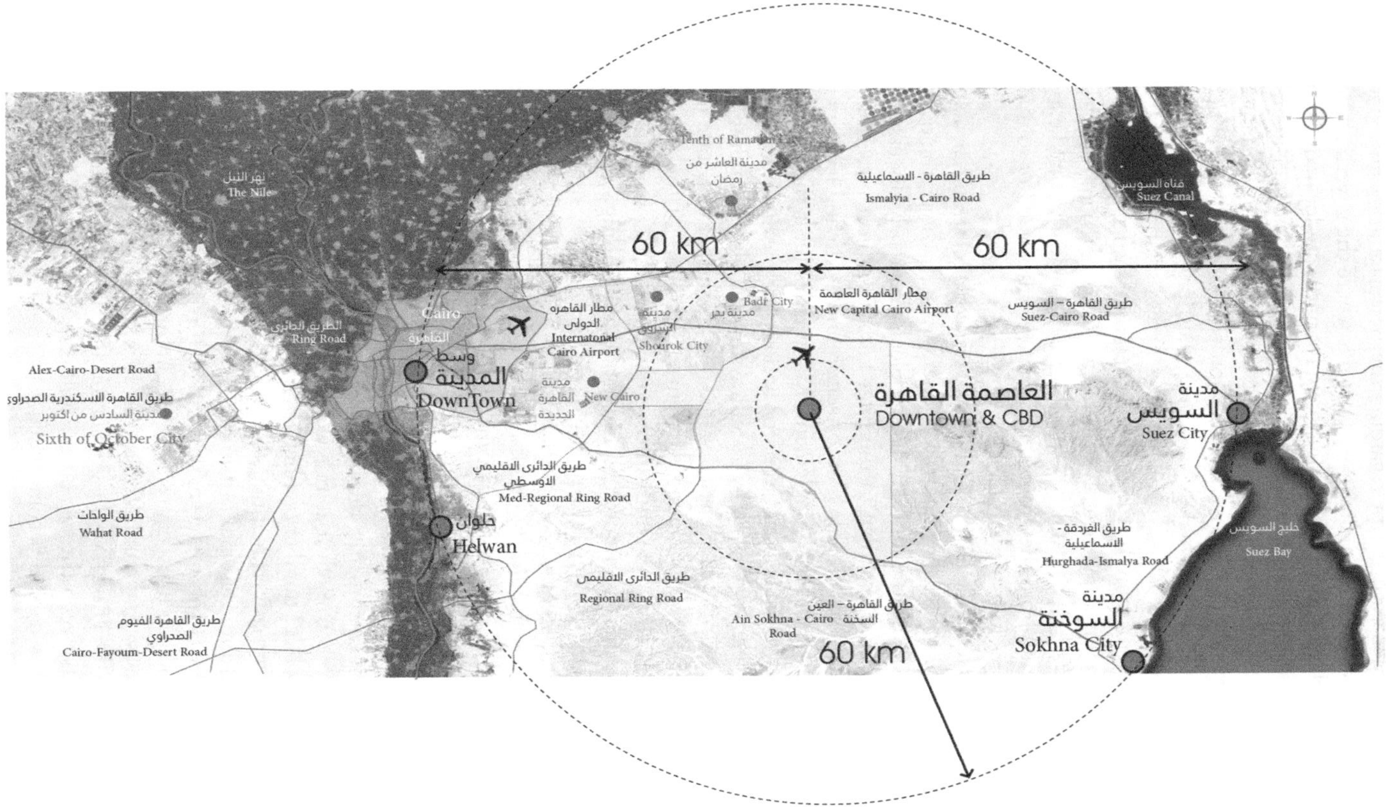

Figure 4.2 The location of the New Administrative Capital

Source: Ministry of Housing, Utilities, and Urban Communities, 2015. http://admin.mhuc.gov.eg//Dynamic_Page/636651596127921042.pdf.

The emerging debate in Egypt regarding the new capital touches on crucial issues, since it asks whether the nation's capital is where the governmental functions occur, or where the heritage, culture, and services are located. The choice confronting decision makers is whether to invest in "old" Cairo or in creating a "new" Cairo. Does the future of Cairo require an investment in the new capital? Or is the main mission of the government to provide for urgent social needs and problems, especially in informal areas? These questions reveal the necessity to establish a wise balance between constructing a new administrative capital and sustaining the current capital. A reasonable compromise suggests that the future of Cairo lies in a clear and distinctive historical city dimension, thus delivering significant added value in the cultural sphere, and in a new structured experience that can be designed from the ground up to deliver sustainability solutions and well-defined services to its new residents.

Every city seeks to achieve the sustainability principles of strong capacity today while preserving resources for tomorrow's generations. However, the overcrowding found in Cairo, and the multilayered problems previously discussed, constitute a great challenge, as shown in the pilot Embaba project presented in this chapter. Delivering a sustainable plan of action and achieving its principles in a new city should be easier to implement. Many cities limit their conceptions of sustainability to specific, uncoordinated actions such as incorporating green roofs and/or graywater systems, or by installing solar panels on building rooftops. However, a new administrative capital would need to further elaborate its design by considering relationships among different components of the city and the core infrastructures, costs of maintenance, and benefits. Most importantly, looking at the hub resources as an interconnected and functioning system of systems will require solutions that provide inclusive and participatory strategies that are sometimes difficult to deploy.

The decision to create a new capital is in line with the vision of "Egypt 2030," (the most recent strategy elaborated by the Ministry of Planning in Egypt) (Egypt Vision 2030, No date). Redirecting investments to new geographies and developing places other than Cairo to strengthen and enhance the economic potential in Egypt through the diversity and competency of different cities is a worthy strategy to explore. Meanwhile, duplication of functions in the role of the new capital could arise if the effort is not well managed. The fact is that a major physical move of a large number of functions and workers is the dominant feature of the plan. Shifting governmental institutes, headquarters of leading companies, national and international organizations, 18 ministries, and all embassies will establish a new government administrative district, a wide variety of urban neighborhoods, and a cultural center. (New Administrative Capital, 2018)

Many more questions abound: Will the creation of a new administrative capital help to reduce the growth of the population in Cairo as expected? Will there be a competition between the two cities? Will Cairo remain the country's main city? Or will the new capital become a major magnet for investments and population? There is no doubt that in order to succeed in achieving a good balance between both cities, there should be a strong governance program linking both cities and their hubs. But such high-level linkages are historically difficult to plan and deploy. The time and the degree to which leadership is open to thoughtful strategy analysis and evaluation will define outcomes – something that is well beyond the scope of the present chapter.

Sustainability as seen by the stakeholders in informal areas

An important debate is going on concerning the New Administrative Capital (NAC) and will continue as the city develops. Practitioners, planners, the media, investors, civil society, and all other stakeholders are organized into two distinct camps: some interpret the creation of the new city as a beneficial initiative, while others criticize the idea of building a new city in the first place, and especially an administrative capital. Social media is playing an important role here, conveying messages of all kinds between supporters and detractors of the NAC. While the government and other diversified stakeholders confirm that this is the right step to revitalize the central core of Cairo, and alleviate the densification, other stakeholders read this action differently, and argue that investments should first prioritize upgrading informal areas which present a great threat to the current capital, and that these areas host a large population that deserve to be considered.

In fact, Cairo's development is not a binary choice, since these two actions perhaps should occur simultaneously in order to preserve the benefits accrued from supporting informal settlements for years, and at the same time allowing some of the capital city hub functions to expand into nearby planned and more predictable service areas.

Although the various new cities that have been built around Cairo have not reached their full capacity yet, we cannot deny that formal extensions and new cities are required, while conceding that they did not reach their full capacity due to many reasons, particularly the lack of public transportation connecting these cities to the capital, and to each other – making it difficult to achieve key sustainability outcomes. Therefore, connecting the NAC to Cairo via a reliable rapid transit system is an absolute necessity. Improving the quality of life in the informal areas of the old Cairo, on the other hand, requires special innovative approaches that can be pursued in parallel to the construction of the new city.

The most relevant discussion here lies in the type of stakeholders that support investment in the new city and those who back concentrating on the old Cairo. Developers, other members of the private sector, and planners, along with the central government, are the main stakeholders interested in the NAC. In contrast, local NGOs, international funding agencies, and academics, along with local government, represent the main stakeholders for interventions in informal areas. Hence, actions are different, stakeholders are different, and the sustainability issues are different: planning a new city offers great opportunities to design sustainable communities, while in informal areas, the challenges are immense, and require not only financial resources, but most importantly, the will of the citizens to participate, support, and sustain the ongoing activities. Despite the different camps and groups of supporters, an effective solution is possible. As Yoon has noted: "do not make a new city to compete with the inner city or mother city. Make it to form a win-win relationship" (Yoon, 2016).

A pilot project carried out in Embaba, an unplanned sector suffering from many problems in the Greater Cairo Region, highlights how residents can become involved in the search for more sustainable futures. If informal areas are to become part of a sustainable capital, their residents have to participate in the planning and service delivery procedures, as the residents of Embaba proved they can in the pilot study described ahead.

The components of the seven-element hub in Figure 1.1 in Chapter 1 of this volume can develop impactful answers to the problems of informal development through the efforts of stakeholders in urban initiatives by means of research (academia), NGOs (small-scale projects), or increased awareness (media). By engaging stakeholders in efforts addressing key shared concerns and with roles supported by new technology, specific weaknesses in the capital city can be addressed and improved. Such an effort to address a priority concern related to water and other key urban resource flows through a technology-based effort was undertaken in 2013. Funded by the Abu Dhabi Global Environmental Data Initiative (AGEDI) and undertaken through a partnership between Ecocity Builders, a U.S.-based NGO, and Cairo University, this project focused on the Embaba neighborhood in Cairo, an informal settlement of significant size and history. The subject focus was water quality and availability, and the technology tools included Geographic Information Systems, smartphones to capture residential-level data, and a system to analyze and present results in easy-to-understand visuals called the Urban Metabolism Information System (UMIS).

Three hub elements were active and central to this pilot effort: academia, NGOs, and civil society with the support of the local government. In order to ensure engagement by the residents, the project invested considerable effort in awareness raising, as well as developing and supporting training from academic experts who volunteered to go into the neighborhood and engage the residents one-on-one.

The government's inability to measure important sustainability indicators in informal settlements in Cairo meant that no systemic solution could be explored. This is why the idea of measuring resource efficiency through technology tools and techniques and using participatory techniques that engaged the population was welcomed by residents and administration alike. The focus of the effort was mainly related to informal areas that have survived over many decades since they have their own pattern of consumption and strong identity, and are densely packed. The public policy question facing these informal areas is not only to define sustainability in a way that makes cultural and historical sense, but to find ways to measure it at a scale that can make a difference.

The initial objectives of the project were to:

- Involve residents using new technologies for data collection and display
- Address the high priority goal of water quality

- Fill governmental informational gaps through resident involvement with household data collection efforts using school children trained by university students
- Provide strong results so that replication and good practice sharing would ensue

The primary outcome of the project was knowledge sharing about how data are being used for smarter, risk-based resource allocation, better sharing of information from agency to agency to facilitate sensible decision-making, and using data in a way that integrates into the established day-to-day patterns of city service providers and citizens. These outcomes were achieved through a pilot effort which developed household-level measurements of water availability and quality.

The conceptual model that was used was that of urban metabolism and resource flows. Tracking a scarce resource such as water through its many uses (and sometimes re-uses) in an urban geography, cataloging both consumption and wastage of water, and attempting to account for all major volumes of water was a notion that was accepted and integrated into the urban planning and research context.

Despite the generally positive results of the project, the problem of funding and financing the implementation of the project outcome was a major issue in seeing rapid implementation. The suggested interventions can be divided into three groups:

- On the neighborhood level, changing the water network in the area, which is a governmental responsibility
- On the buildings and units level, replacing deteriorated water pipes, adding water filters, and installing graywater systems
- On the individual level, promoting efficient use of water through behavioral awareness and education (Attia & Khalil, 2015)

There was some evidence of the capital city hub working throughout this pilot study. For example, at Cairo University, relevant resources were integrated into the course syllabus, while select seminars and labs that complemented course objectives were offered in accordance with the project's needs. In this way, on-the-ground observations and civil society observations influenced the academic offerings, showcasing synergy across the hub.

A strong reason for selecting Cairo locales for the pilot was the expectation that, if the project works in the capital, local authorities in other areas who see innovation would appreciate it and promote its adoption throughout the country. Although it is hard to assess whether this in fact happened, the research team participated in many outreach programs, right down to local news programs, increasing the chance of innovation dissemination. The process of the NGOs' engagement, and the improved understanding of resource flows in the informal communities, are both vital to the success of any government program, and propel forward the expectation that, through effective communication programs, the strategy might be adopted by other actors within and across the hub.

There is a hope that there will be more similar actions to engage civil society at the granular level of the neighborhood, and that the government as the main actor responsible for developing national strategies and policies can find simple, low-cost, and effective ways to engage civil society and respond to people's need for a better quality of life.

It is also true that, even if there is a new capital, and there is a plan for deconcentration, many living communities will remain behind in the "old Cairo" – and their voices will still need to be included in national policies. The concept of urban metabolism and citizen engagement will be there to make a difference under both scenarios.

The future of the Greater Cairo Region: vision versus challenges

Ultimately, a new capital means moving both functions and residents to a new location, which brings to the surface one of the most difficult questions: Will Cairo achieve any benefits from moving some parts of the capital hub functions to a new city? The new capital is currently under construction, accelerating the vital question of hub deployment and survival.

As this discussion shows, different parts of the hub favor the old and new Cairo. The private sector, planning community, and central government are the key stakeholders interested in building the new city, while local NGOs, international funding agencies, and academics, along with local government, have a strong interest in improving conditions in the informal developments of old Cairo. Many scenarios may be foreseen in the future operation of the hub as Cairo continues to evolve:

- One optimistic scenario is that the hub systems will be duplicated in both cities, and that this duality will positively affect the efficiency of the cities' functioning and achieve a win-win relationship. In this case, elements of each hub system will be well connected, governed, and financed. Such a successful outcome will depend strongly on the connectivity of the current capital and the NAC, stressing three principles:
    - Inclusive planning
    - Access to public transportation
    - Affordable housing for all
- A second alternative scenario suggests that Cairo will retain the major hub functions, and the administrative city will act as an extension of Cairo offering high-end communities, and investments in services. This scenario would copy previous urban extensions and is likely to happen if the three principles from the first scenario are not achieved.
- A third scenario implies a strong development of the new administrative capital, where Cairo will be left behind with minor investments, and will host more informal developments and businesses. However, the history of

Cairo dictates that it will remain the main capital city hub at least for the next 30 years. It has been a resilient city facing political, economic, and social challenges over centuries, but this will not prevent further deterioration, unless a strong management system, efficient governance, and the participation of all stakeholders towards achieving sustainable development is assured, making the Embaba type of engagement vital to success.

Only time will tell which future becomes reality. A successful reality requires three major tools:

- Pursuing an effective governance strategy; if leaders understand the hub linkages and their importance, one can expect good results and efficient government programs.
- A large campaign of capacity building and training for local government officials, and other stakeholders; these capacities include on-the-job training, innovative approaches, and contextualized systems relevant to the human resources which are available, while improving their knowledge, culture, and coordination and management skills.
- Developing strong regulations linking urban development to investments and services provision.

The hope of the authors is that experiences from Cairo, from older generations of new cities in Egypt, and from the other capitals in the articles of this volume will stimulate deeper research and experimentation that will lead to stronger strategies and clearer pathways to sustainability for capital cities. The search for "the" unique and clear solution will continue to prove elusive. What is sure is that the road to discovery must be taken at once, and that residents, be they from planned or informal sections of capital cities, must engage in the processes of planning, representation, and action using tools that make such participation feasible for all.

Note

1 In Arabic spelled "Al Qahera," The victorious.

References

E. Ali. (2003), Evaluation of the Egyptian Experiment in Establishing the New Towns in the Desert Areas. *Journal of Engineering Sciences* 31 (1): 231–244.

Attia, S. (2012). *Revitalization of Downtown as Center for Social Democracy and Sustainable Growth: Ecocity Builders Summit Book*. Montreal: Ecocity Builders.

Attia, S. and Khalil, H. (2015). *Urban Metabolism and Quality of Life in Informal Areas*. Proceedings of the 20th International Conference on Urban planning and Regional Development in the Information Society. Real-Corp. Plan Together-Right Now-Overall, 661–674.

CAPMAS. (2017a). *CensusInfo*. Retrieved from the Central Agency for Public Mobilization and Statistics: www.capmas.gov.eg/Pages/ShowPDF.aspx?page_id=www.censusinfo.capmas.gov.eg/

CAPMAS. (2017b). *Final Results of Census of Egypt*. Retrieved from the Central Agency for Public Mobilization and Statistics: www.capmas.gov.eg/Pages/ShowPDF.aspx?page_id=/Admin/Pages%20Files/Presentation.pdf

CAPMAS Censuses. (2017c and 2006). Retrieved from the Central Agency for Public Mobilization and Statistics: www.capmas.gov.eg/Pages/IndicatorsPage.aspx?page_id=6156&ind_id=4575

Egypt Vision 2030. No date. http://sdsegypt2030.com/?lang=en (accessed January 3, 2019.

GOPP (2012). *Greater Cairo Urban Development Strategy, Part I: Future Vision and Strategic Directions*. Cairo: The Ministry of Housing, Utilities, and Urban Communities (MHUC) represented by the General Organization for Physical Planning (GOPP).

Hall, P. and Pfeiffer, U. (2000). *Urban Future 21: A Global Agenda for Twenty-First Century Cities*. Oxon: Taylor & Francis.

Hegazy, I. and Moustafa, W. (2013). Toward Revitalization of New Towns in Egypt Case Study: Sixth of October. *International Journal of Sustainable Built Environment* 2 (1): 10–18.

Ministry of Housing, Utilities, and Urban Communities. (2014). *The National Urban Development framework in the Arab Republic of Egypt*. Cairo: Ministry of Housing, Utilities, and Urban Communities. http://gopp.gov.eg/wp-content/uploads/2015/01/GOPP-PA_2014.pdf. Accessed January 8, 2019.

Ministry of Housing, Utilities, and Urban Communities. (2015). Untitled map. http://admin.mhuc.gov.eg//Dynamic_Page/636651596127921042.pdf

New Administrative Capital. (2018, March 8). Retrieved from State Information Service: www.sis.gov.eg/section/352/5238?lang=en-us

Provoost, M. (2016). *Achieving Sustainable New Town Development in Developing Countries*. Quito, Equador: UN Habitat.

Sims, D. (2003). *Understanding Slums: Case Studies for the Global Report on Human Settlements*. Cairo. Retrieved from www.ucl.ac.uk/dpu-projects/Global_Report/cities/cairo.htm

UN Habitat. 2011. *Cairo: A City in Transition*. Nairobi: United Nations Human Settlements Program.

Werthmann, C. and Bridger, J. (2015). *Metropolis Nonformal*. San Francisco: Applied Research and Design Publishing.

Yoon, Jungjoong. (2016). *Achieving Sustainable New Town Development in Developing Countries*. Quito, Equador: UN Habitat.

5 Stockholm

Where sustainability meets technology

Anna Helm and Nicole Forsberg
with Christina Johannsen

In recent decades, Stockholm has developed a strong, innovative economy with an impressive focus on environmental sustainability, even earning the distinction in 2010 as the world's first European Green Capital (European Commission, 2018). Stockholm's path from a remote, polluted, unsophisticated capital with an economy based on raw materials to a globally recognized leader in sustainability with a dominant service sector can be largely explained by the symbiotic and mutually supportive relationship between the government, the private sector, and academia.

The goal of this chapter is to explore this dynamic system of relationships through the lens of the Hub Theory. This chapter will lay out the key accomplishments in sustainability and innovation that have transformed Stockholm into a world leader in sustainability innovation. Several historically significant developments that contributed to the status of Stockholm as a notable stronghold for sustainable economic growth will be described. Four unique components, or "spokes," of the hub described in Chapter 1 of this volume will be at the center of the discussion: the city government, the national government, research communities, and corporations. Each of these spokes has played a distinct role in the development of Stockholm into a world-renowned center for sustainability innovation. Subsequently, an explanation of how these actors link together in the capital city to produce these innovative results will be provided. Further, this chapter will present limitations to the Hub Theory in explaining Stockholm's success in sustainability innovation.

Stockholm provides a compelling example that national capital cities can, in fact, leverage the various spokes of the hub in order to push the frontier of urban sustainability and innovation forward as hypothesized in Chapter 1 of this volume. This chapter draws on analysis from published data and interviews conducted with key opinion leaders within the Stockholm innovation and sustainability arena to support its arguments.

Key accomplishments

Sweden, a country situated in the Nordics with roughly 10 million inhabitants, is frequently mentioned in discussions about sustainability innovation. In a 2015

Country Sustainability Ranking which aimed to map out the most relevant risk and return drivers for investors, Sweden was ranked the "Most Sustainable Country in the World" out of 59 countries based on 17 indicators in the environmental, economic, and financial governance areas (Robecosam, 2015). This is not a new phenomenon; Sweden has consistently been considered an international frontrunner in sustainability. The *Naturskyddsföreningen*, or the Swedish Society for Nature Conservation, was established in 1909; in 1967, Sweden became the first country in the world to implement an environmental protection agency; a separate Ministry of the Environment was established in 1987; and already in 2012, Sweden met its 2020 goals for renewable energy (Sweden.se, 2018a).

Stockholm, the largest city in the Nordics, is the capital city of Sweden with roughly 2.3 million inhabitants in the metropolitan area as of 2016 (SCB, 2017). The city is located on the east coast of Sweden and is made up of a conglomeration of islands. The Stockholm archipelago, starting just a 20-minute drive from the city center, consists of an astonishing 30,000 such islands, which are attractive to Stockholmers as vacation destinations during the summer months.

As evidenced in the passion its citizens have for nature, including these islands, Stockholm has a historic reputation as a beacon in sustainability innovation. This status can be traced back to 1972 when the city hosted the first United Nations Conference on the Human Environment (United Nations, 2018). In 2010, Stockholm was designated the first European Green Capital by the European Commission, and according to Björn Hugosson, Head of Climate for Stockholm City, the prestigious award set a high bar for Stockholm to continue to grow as one of the world's leading cities in urban sustainability (Hugosson, 2018).

It is important to note that the initial industrial growth in Stockholm was based on raw materials, but later this growth was attributable to an innovation- and science-based industry led by globally renowned Swedish inventors and entrepreneurs such as Lars Magnus Ericsson, Alfred Nobel, Gustaf de Laval, Aron Andersson, and the Wallenberg family. As host to these burgeoning corporations, Stockholm became more attractive to the international business and research community and began to leverage its status to import foreign skills and knowledge. Along with these technological developments, the city began to modernize its infrastructure, and citizen quality of life grew as new projects like water piping and sanitation developments were brought online (Metzger et al., 2013).

Stockholm has been recognized as a top tier city for technology innovation. It is the birthplace of global technology firms such as Ericsson and Atlas Copco, and in recent years has attracted increased attention for technology innovation within the Information and Communication Technology (ICT) sector as the city of origin for fast growth technology companies such as Skype and Spotify. As a result of the increased attention these prominent technology companies have drawn to Stockholm, the city can claim to be one of the most powerful and dynamic technology clusters in the world. Such prowess is evidenced by the volume per capita of "unicorns" found in Stockholm. Unicorns, originally defined by venture capitalist Aileen Lee, are startup companies with a valuation in excess of US$1 billion (Lee, 2013). As of December 2017, there are 268 worldwide (Atomico, 2017), and of the nine Unicorns founded in the Nordic countries, six are in Stockholm. In fact, Stockholm ranks second in the world per capita (with a total of 6.3 Unicorns) for these successful ventures, second only to Silicon Valley's 6.9 (Davidson, 2015).

Stockholm has combined environmental sustainability with high economic growth, fueled by new technology innovations with global impact. This chapter will explore Stockholm as a center for sustainability innovation by focusing on the connections between relevant spokes of the hub, tracing important historical milestones that set the tone for the dynamic development of Stockholm into a capital city of prominence in the realm of sustainability innovation.

Spokes of the hub

The national government, city government, local universities and researchers, and private sector all play key roles in the development of sustainability policies and practices in Stockholm.

National government

The national Swedish government has established ambitious sustainability targets. Sweden's "Roadmap 2050" mandates the elimination of net GHG emissions by 2050, including sub-goals such as the electrification of the country's vehicle fleet by 2030. This target, however, requires significant investments. In 2016, the Swedish government set aside 4.4 billion SEK (approximately $500 million) toward renewable energy and infrastructure projects like smart grids, solar energy implementation, and clean transportation. In keeping with the intent of the investment, much of the funding for this initiative came directly from heavy carbon taxes on non-renewable emission sources like diesel and gasoline, as well as the sale of coal mines (Sweden.se, 2018b). Forward-thinking actions and policies like these from the national government have had, and will continue to have, a positive impact on sustainability within the capital city. Similarly, Stockholm's reputation as a premier international innovation hub is driven in large part both by policies enacted at the Swedish national government level and the derivative policies at the city government level.

District heating

The Swedish government has been supportive of the implementation of district heating systems since the 1950s. District heating is generated from recovered waste heat from combined heat and power plants, industrial processes, or waste incineration and distributed to customers as hot water through a pipe network. Aside from being an efficient and comparatively inexpensive way of delivering heat to the residents of a city, this method of heating replaced oil-fired burners, which had significantly greater negative environmental impacts. Beyond the environmental benefits, the financial advantages of district heating were especially evident during the 1973 oil crisis. After the successful implementation of district heating in Stockholm and other major cities in Sweden, the development expanded to various towns within the country through the 1980s (Ericsson et al., 2016). Today, district heating is the most common heating source in Sweden; in fact, approximately 90 percent of apartments in Stockholm now use it as their primary heating method (Svebio, 2018). The implementation of district heating since the 1950s has contributed significantly to the reduction of greenhouse gas (GHG) emissions in Stockholm (OECD, 2013), as the system in Stockholm is largely fueled by biomass and waste from the city's inhabitants, along with a small amount of fossil fuels. Hugosson (2018) reinforces the economic and environmental sustainability of district heating as a reflection of the active resourcefulness of Stockholm citizens – supported by the local government – in using an otherwise wasted resource: heat as a byproduct of electricity production.

Hydroelectric power

Renewable sources of electricity generation have been utilized in Stockholm since the end of World War I. The Swedish government, for example, has invested heavily in hydroelectric power as a major renewable energy source in Sweden. Stockholm's proximity to water has naturally led the city to rely on hydroelectricity, which is especially attractive because of its relatively low impact on the environment.

Carbon tax

The "polluter pays" principle – where the costs of pollution should be shouldered by its emitter – was developed to promote the reduction of pollutants in a cost-efficient manner, and to incentivize the utilization and development of new innovative and pollution-reducing technologies. This principle, and the promise of resulting innovation, is the driving force behind the carbon (CO_2) tax introduced in Sweden in 1991, aimed at increasing the amount of renewables in the Swedish energy mix. While energy sources in Sweden have been taxed since the 1920s, this carbon tax was implemented as an addition to a pre-existing energy tax and has significantly contributed to the reduction of carbon emissions in Sweden and to the city of Stockholm since its introduction. This beneficial change is especially prominent within the residential heating sector, where the increase of biomass usage rose from 25 percent in 1990 – just prior to the tax – to 70 percent in 2012. According to Swedish finance minister Magdalena Andersson, in an article written for the World Bank, the carbon tax has also stimulated growth of new technologies that are developed by the private sector. Andersson also states that putting a price on carbon is not only a morally sound thing to do, it is also smart politics and relatively easy to implement (Andersson et al., 2015).

The carbon tax generates a considerable amount of money for the general budget of the country, and according to the Swedish government, it has been relatively easy and cost-efficient to implement. In terms of measuring emissions for taxation purposes:

> [T]he carbon tax is levied on all fossil fuels in proportion to their carbon content, as carbon dioxide emissions released in burning any fossil fuels are proportional to the carbon content of the fuel. It is therefore not necessary to measure actual emissions, which greatly simplifies the system.
>
> (Government Offices of Sweden, 2018)

Investments in new technology

The Swedish government has historically been a prominent investor in new technology. An example of this effort is Industrifonden, which was initially started and funded by the Swedish government in 1979, and is today an independent evergreen venture capital firm which runs independently and reinvests all profits into new companies. Through the years, Industrifonden has funded hundreds of new technology-based ventures in Sweden (Regeringen, 2018a), but without initial government support would not have been able to thrive. Other governmental investment companies include Fouriertransform, which has a strong focus on innovations within the manufacturing sector and automotive industries, and Inlandsinnovation, which focuses on innovations in northern Sweden. In March 2016, the Minister for Enterprise and Innovation ordered government-funded venture capital firms to increase their efficiency (Regeringen, 2016). Saminvest, started directly in response to this order, is also government funded and invests "indirectly in companies by co-investing in venture capital funds in areas where there is a need for complementary market investments" (Saminvest, 2018).

Similarly, Almi Företagspartner AB is owned by the Swedish government and is the parent company of a group consisting of 16 regional subsidiaries. Almi provides financing, business support, and loans for companies or individuals wanting to grow their business or turn an idea into a viable company. For Swedish citizens, Almi support is free of charge, and is a service aiming to drive entrepreneurship and innovation among Swedes. Almi also has a venture capital arm, Almi Invest, which is the most active venture capital investor in Sweden. Initially financed by the government, it now reinvests all profits into new companies, similar to Industrifonden. In 2016, Almi Invest launched its GreenTech Fund with the purpose of "bridging the market gap that exists between demand and supply on venture capital investments for innovative corporations contributing to the reduction of greenhouse gas emissions" (Almi, 2018).

Like previously mentioned government-funded efforts, Vinnova is a government agency in the arena of research and innovation. The vision of Vinnova is to help Sweden "become a leading global player in research and innovation, and a country that is attractive for investment and entrepreneurship" (Vinnova, 2018a). Vinnova has played an important part in the funding of research and innovation projects, specifically for proof-of-concept ideas and those in the early stages of innovation to help reduce ventures where "the risks are great and . . . projects would generally not get off the ground without government aid" (Vinnova, 2018a). For the Stockholm area, support from Vinnova has contributed significantly to the creation of new ventures from local university labs, according to Lisa Ericsson, Head of KTH Innovation, the innovation and technology transfer unit at KTH Royal Institute of Technology, the largest engineering university in Sweden, located in Stockholm (Ericsson, 2018).

With this new approach to government-funded venture capital, Sweden looks to strengthen the capital supplied to high-risk investments. Government-backed firms like Saminvest, by co-investing with private capital, and government agencies like Vinnova, through risk reduction and early funding of innovation, aim to develop the venture capital market in Sweden as a whole, ultimately strengthening the country's climate for innovation and economic growth.

Computer subsidies

In 1998, the Swedish government enacted a personal computer subsidy at the national level which allowed employees to lease personal computers tax-free for three years, after which they were allowed to buy them at the market price. This is perhaps the most effective governmental policy driving innovation among Swedes, as it "had a huge impact on computer literacy, and – coupled with the harsh Swedish winters, which tend to keep families indoors – meant that a generation of children grew up with impressive computer skills" (Index Ventures, 2016). According to Marie Wall, Deputy Director-Startups, Swedish Ministry of Enterprise, the policy allowed many people who might not have been able to afford a computer to buy one, which ultimately brought early computer literacy to the general public. Approximately two decades later, the benefits are evident as Stockholm has turned into one of the world's most prominent technology hubs where the most common job title found in Stockholm today is "programmer."

> This type of talent is central in the building of an innovation hub – and the personal computer subsidy has had a direct effect on developing Stockholm's computer science talent pool, infusing computer-literate professionals into the city as well as professionals who have grown and innovated within the hardware, software, and digital design communities there.
>
> (Wall, 2017)

Undoubtedly, the computer subsidies have also contributed to the volume of technology-based unicorns born in Stockholm, as well as helping to drive innovation and technology forward in large multinational corporations like Ericsson and Kinnevik. The now-inherent interest in technology has also driven growth for technically focused educational institutions in Stockholm like the KTH Royal Institute of Technology. KTH has been an incubator for a majority of the Stockholm-based unicorns and is often referred to as the "MIT of Stockholm" by the international media (Cohan, 2017).

Innovative Sweden

In 2004, "Innovative Sweden" – the first national innovation strategy for the country – was published (Regeringen, 2004). This report gives a comprehensive overview of the strategies that are prioritized by the nation to promote long-term sustainable growth. This growth is projected to occur as a result of cutting-edge technology and innovation, and is expected to strengthen Sweden as a knowledge-based economy. The report lists education and research as priorities, since they play an "ever-increasing role in economic growth." The vision put forth in this document is for "Sweden to be Europe's most competitive, dynamic and knowledge-based economy, and thus one of the world's most attractive countries for investment by large and small knowledge-based companies." Country growth and sustained international competitiveness are also acknowledged to be a result of the unique and historically strong welfare system. In Sweden, it is widely believed that a "well-functioning social security system, combined with good opportunities for skills development, increases the prospects of achieving change without excluding significant groups from the labour market." Interestingly, the "Innovative Sweden" strategy acknowledges the importance of keeping its public sector lean, innovative, and evolving with new technology development, even though the government employs roughly thirty percent of the workforce. Further developed in this strategy is the importance of the interconnectivity and collaboration between the public and private sectors, which contributes to the development and up-scaling of new technologies, especially in telecommunication, energy, and transportation.

City government

Through the Green Cities Program, a case study on Stockholm as the world's leading international city was published in 2013, aiming to highlight best practices in sustainable urban development (OECD, 2013). The report examined the policies, strategies, and actions taken by the government that have led to the city's sustainability successes. Stockholm was chosen as the initial case study because its:

> population and economic growth are well above the OECD average. In the past decade, the service sector has been Stockholm's fastest-growing economic sector, and this economic success has been matched by strong environmental performance. CO_2 emissions per capita are among the lowest of any OECD metro area, water quality is excellent, very little waste goes to landfill, and air pollution is below critical measures for almost all pollutants. (OECD, 2013)

Government role in alternative energy innovation

Stockholm is traditionally an early adopter within Sweden for innovative solutions in alternative energy, which has driven it towards successes in urban sustainability. This energy focus can be seen specifically in the previously discussed cases of district heating and hydroelectric power, as well as through government initiatives in terms of taxation (dis)incentives. Stockholm therefore has reaped the benefits, both economic and social, of this early adoption trend, contributing to a resource- and cost-effective system (OECD, 2013).

Stockholm pushed the boundary on sustainable initiatives, adopting its own "Strategy 2040" which aims to achieve the goals set for Sweden ten years earlier than planned within the city limits. These ambitious goals of Stockholm will be met through encouraging local development and implementation of sustainable energy consumption, environmentally friendly transportation, and efficient recycling. It is even committing to be – as a city government organization – fossil fuel free by 2030.

Stockholm's bid to host the 2004 Summer Olympics was an important milestone in the city's rise to becoming a frontrunner in sustainability. Inspired by Sydney's 2000 branding as the "Green Olympics," Stockholm decided to transform the Hammarby Sjöstad district into an eco-friendly site for an Olympic Village. At the time, Hammarby Sjöstad was an industrial waterfront site, and these plans would reconstruct it into a state-of-the-art sustainable district with "previously unheard of environmental standards" (Metzger et al., 2013). While Stockholm lost the Olympics bid to Athens, the city government decided to continue with construction of the flagship project, integrating into it the most cutting-edge ideas in green technology concepts and sustainable urban planning. This investment has paid off. Since completion, the area has flourished as both an attractive residential neighborhood and as an international showcase for Swedish green technology – an important part in the country's export strategy (Metzger et al., 2013).

Stockholm's governmental commitment to sustainability and its realization that innovation is a central part of driving sustainable urban development are key to the performance seen today in the city. The work done to deserve – and the data collected to validate – the award as the first European Green Capital was, as Hugosson (2018) noted, "a great way to visualize the efforts that had, and were ongoing, in the city in regard to sustainability." While this award was an honor, Stockholm realized that continued innovation to build upon previous successes would be integral to meeting its ambitious city and country targets. These targets and self-made challenges can be met through support from the city's innovation strategy aimed to create an effective and creative collaboration between the different actors in the city (Hugosson, 2018). In proclaiming these collaborative efforts, Stockholm has identified the primary actors of the Hub Theory as integral to its success as a leader in sustainability innovation.

The subway system

As Stockholm's economy transitioned from a reliance on the extraction of raw materials to technology and innovation, manufacturing operations moved to rural areas surrounding the city. Management and research and development (R&D)

operations became the focus in Stockholm's city center for large companies. As a result, other services-based companies, notably the consulting and creative industries, began to also concentrate operations in Stockholm's city center, setting the foundation for the city's present-day layout. Concurrent with these technological advances, Stockholm's development of a subway system in 1944 served to connect these industries, locations, and employees (Metzger et al., 2013). At the time, Stockholm was "probably the smallest city in the world implementing a subway system" (Gullberg, 2017). The population of the city was moderate in size and subway construction was therefore a substantial economic and strategic investment by the city. Today, using the subway is often the fastest and most convenient and cost-effective way to get around Stockholm and its suburbs, and is the preferred mode of transportation for a large portion of the inhabitants in their daily commute. That it is an environmentally friendly choice is simply a bonus. Stockholm has accordingly invested in the commuter journey, and made the subway experience exciting, as the Stockholm subway system also is said to be the world's longest art exhibit. "Over 90 of the 100 subway stations in Stockholm have been decorated with sculptures, mosaics, paintings, installations, engravings, and reliefs by over 150 artists" and at 110 kilometers long, traveling through the subway system "is like traveling through an exciting story that extends from the artistic pioneers of the 1950s to the art experiments of today" (Visit Stockholm, 2018).

Government taxation and incentives

In terms of alternative energy innovation, greenhouse gas reductions in Sweden have also been propelled by the introduction of Stockholm's congestion taxes, which were enacted in 2007. The taxes incentivize city inhabitants to use environmentally friendly transportation rather than personal vehicles in an effort to reduce emissions. The revenue from these congestion taxes is directed towards the development of public transportation and the maintenance of roads in the city.

A networked city

In 1994, as computer literacy was increasing, and just prior to the personal computer subsidies, Stockholm launched Stokab, "a municipally owned passive fiber infrastructure provider . . . to lower its network costs, increase competition for telecom services in the city, and minimize disruption to citizens by limiting private operators' needs for civil works to lay infrastructure in the ground" (Felten, 2015). Stokab was billed as a public infrastructure company by the City Council and was originally financed by Stockholm to connect public institutions and universities. It began deploying to residential buildings in 2002, and in present-day Stockholm, there are now over 30 different options for home internet service providers and 120 providers for businesses, and over 95 percent of all homes have access to the network (Index Ventures, 2016). This network – and the resulting quality of telecommunications services – has been a key factor in business attractiveness for the city. In the Cushman & Wakefield annual reviews of the perceived attractiveness of European cities, Stockholm has consistently been ranked in the top four cities for the quality of telecommunications, typically only behind London, Paris, and Frankfurt (since 2001, excluding 2007). While this connectivity is not the sole reason for Stockholm becoming the innovation hub that it is today, "the presence of a state-of-the-art IT infrastructure that enables competition and therefore affordable high-quality services is, alongside other factors such as available tech talent and emerging venture capital initiatives, [one contributing reason] why Stockholm has become such a hub for tech innovation" (Felten, 2015).

Universities and research institutions

As one of the core pillars for creating successful innovation hubs (Wall, 2017), talent fostered within the academic sphere in Stockholm is integral to the city's development and successful performance in sustainability and innovation. Several prominent educational institutions are located in Stockholm: the KTH Royal Institute of Technology, Sweden's largest technical and engineering university; Stockholm University; Karolinska Institutet, one of Sweden's foremost medical universities; and the Stockholm School of Economics, Sweden's top-ranked business school. Aside from the larger universities there are also a cadre of specialized schools like the Royal College of Music; The University of Arts, Crafts and Design (Konstfack); and the Berghs School of Communication that complement Stockholm's academic strength in business and technology. The variety of academic focal areas provides Stockholm with a deep bench of talent. In fact, in a report released by Fölster et al., 2017, Stockholm ranked second in Europe among capital cities in "brain business jobs," or rather "jobs that are crucial for income and productivity growth," and Sweden – as a country – ranked first. According to Fölster, "a high concentration of brain businesses is a breeding ground for creative ideas, R&D inventions, innovative business concepts, business ventures, and start-ups," which further substantiates academia's contributions to the success of Stockholm as a hub for innovation.

The Swedish government invests heavily in R&D and allocates a large part of its total GDP – 3.25% compared to the OECD average of 2.55% in 2015 – to the funding of research at these Swedish academic institutions (World Bank, 2015). In Sweden, most R&D funding comes from Vetenskapsrådet, which focuses on natural sciences, technology, medicine, and social sciences, while R&D funding for innovation activities is funneled through Vinnova (Regeringen, 2018b). This large outlay of government expenditure and prioritization of R&D drives innovation – specifically in sustainability – in academia, and is one of Stockholm's key strengths as a successful innovation hub (Ericsson, 2018).

Funding of R&D in Sweden is not limited to government sources. Of the many institutions that support R&D, the Swedish Foundation for Strategic Environmental Research (MISTRA), in particular, focuses on sustainability and innovation and supports cross-disciplinary research to solve complex environmental issues in society. MISTRA funds projects that are at the intersection of academia, government, and the private sector, and stresses that collaborative research is integral to solving environmental issues in society. For example, the MISTRA Sustainable Accessibility and Mobility Services (SAMS) research project is housed within and run by KTH in collaboration with the Swedish National Road and Transport Research Institute (VTI). This project focuses on the development of sustainable mobility in an urban environment through the integration of feedback and research from partners within academia, the private sector, and the government.

Professor's Privilege

Since the 1950s, Sweden has had a unique policy – "Lärarundantaget," or "Professor's Privilege" – which permits professors, and not the universities they work for, to own their inventions and research results (the intellectual property created). This system of intellectual property rights creates an ability, and is a strong incentive, for individual researchers to leverage their inventions and technology developments to create spin-off companies from universities. Within KTH Innovation, for instance, there are hundreds of researchers who have taken advantage of the policy. While there has been some question as to whether or not Lärarundantaget is fruitful for the system as a whole, as long as there is a structure in place at the university to handle and capture the intellectual property, it can be very effective, according to Ericsson (2018). Monica Ek, Professor of Wood Chemistry and Head of the Department of Wood Chemistry and Pulp Technology at the KTH Royal Institute of Technology, noted that Professor's Privilege has enabled the commercialization of several groundbreaking technologies developed in the School of Engineering Sciences in Chemistry, Biotechnology and Health. Professor Ek took advantage of it herself when patenting a technology that enables cost-efficient production of nanocrystalline cellulose, which can be applied as an environmentally friendly substitute in various applications, from makeup and paint to packaging materials. She believes that without the Professor's Privilege, the incentive may not have been strong enough to move forward with commercialization (Ek, 2018).

Historically, European universities exist within the "academic innovation paradox" (Atlantic Council, 2018) whereby a significant amount of money is spent on research that does not necessarily lead to significant societal benefit, mostly because its end goal was research for its own sake. In Sweden, however, the "Third Mission" calls upon universities to deliver innovation technology transfer (Atlantic Council, 2018) beyond the already established deliverables: research and education. In conjunction with the new mandate, major Swedish academic institutions set up holding companies tied to the university's investment in research spinoffs from their respective departments. The introduction of the Third Mission for universities in Sweden has led to the birth of several successful innovation units producing multiple new research-based ventures every year (Ericsson, 2018).

Private sector

The private sector represents the last spoke of critical importance to Stockholm's development into a central geographic area for sustainability and innovation. Swedish industry developed relatively late in comparison to the rest of Europe (Metzger et al., 2013), but once the transformation from agrarian to industrial society started, a few key inventions propelled the growth. These inventions laid the foundation for what later became multinational corporations (MNCs) with pervasive and powerful networks surrounding them, along with clusters connected to the core competencies of those MNCs. These clusters fostered talent and gave rise to expansive new companies, whose founders and management teams were interested in giving back to the next generation of entrepreneurs through financial and intellectual investments. This so-called recycling effect (Wall, 2017) has propelled the growth of new companies in Stockholm.

Two prominent individuals spurred Stockholm's emergence as an industrial powerhouse: Lars Magnus Ericsson and A.O. Wallenberg. The former was the founder of telecom giant Ericsson in the late 1800s, and was instrumental in the development of telecommunications on an international scale. The latter left his career as a naval officer – also in the late 1800s – to capitalize on the new wave of entrepreneurship in Sweden, going on to become a major player in the reform of the Swedish banking system. Stockholm's technology cluster emerged around Ericsson, which is today one of the world's leading information and communication technology providers, and has since increasingly focused on ICT, helping build global appeal and international participation through the presence of companies like IBM and Microsoft.

Around the same time as the rise of Ericsson, in 1856 Stockholm's Enskilda Bank (SEB) was founded by A.O. Wallenberg. Now one of Scandinavia's largest banks, at the time of its founding SEB was Stockholm's first private bank and one of the first commercial banks in Sweden (SEB, 2018). It provided local firms in Stockholm the ability to access capital and allowed them to grow at rates that would otherwise not have been possible. The Wallenberg family has since contributed significantly to the development of business and research in both Stockholm, Sweden, and abroad. Through Wallenberg investment companies, such as Investor AB, the family holds stakes in many of Sweden's largest and most prominent corporations, including ABB, AstraZeneca, Atlas Copco, Electrolux, Ericsson, SAAB, Scandinavian Airlines, and Stora Enso (Milne, 2017).

An important factor contributing to the success of Stockholm as a technology hub, which is especially prominent in the private sector is the "intrapreneurship" within the large Swedish corporations, a deeply embedded part of the culture. Pär Hedberg, CEO at Stockholm Innovation and Growth (STING), argues that one of the main factors behind the success of Stockholm and its progress as an innovation hub from the private sector perspective, is the cultural conditioning in Swedish corporations; "from early ages Swedish children learn to adopt an independent mindset, to collaborate, and to question things – a critical and collaborative mindset that we see among Swedish corporations today" (Hedberg, 2018).

This independence and non-authority dependent behavior Hedberg refers to has contributed to the strong reputation Swedish engineers have attained in the global community:

> Swedish engineers are known for their collaborative mindset and creativity – being able to develop complicated solutions without specifications . . . although there is prestige, in comparison to many other countries there is very little of it and the collaborative mindset we see is truly an asset to productivity and innovation.

Another part of the puzzle, according to Hedberg, is the citizen trust in the Swedish government:

> In Sweden we trust the government, and that, in combination with a large and strong government, can enable the country to be a frontrunner in new areas such as digitalization – we see several of our public services being digitized at a much faster pace than many countries in the world. This naturally increases the government sector's purchasing power for new technologies and innovations developed at Swedish corporations.

These factors improve Sweden's position within Hofstede's cultural dimensions in terms of hierarchy. Two of these factors play important roles in the status of Sweden – and Stockholm in particular – as an innovation hub based in part on the private sector. The first of these, the Power Distance Index (PDI), indicates "the extent to which the less powerful members of institutions and organizations within a country expect and accept that power is distributed unequally" (Hofstede-Insights, on-line database). Simply put, a low PDI for Sweden (a score of 31) means that power is distributed more equally, hierarchy is primarily a formality, and that ideas from team-members are solicited and valued as much as those of team leaders or managers. The second important dimension of note is Uncertainty Avoidance (UA), which indicates "the extent to which the members of a culture feel threatened by ambiguous or unknown situations and have created beliefs and institutions that try to avoid these" (Hofstede-Insights). Sweden also scored low here – 29 – which indicates that its citizens typically are open to new ideas, and most importantly that innovation is non-threatening, and the result of this can be seen in the "intrapreneurship" culture in corporations in Stockholm. Together, Sweden's low scores of 31 in terms of PDI and 29 in UA indicate that Stockholm corporations have a natural urge for innovation.

The legacy of the multinational corporations in Stockholm, such as Ericsson and Kinnevik, manifests itself in today's entrepreneurs, many of whom were groomed in these organizations (Wall, 2017), and who continue to invest in Stockholm's commercial ecosystem and thereby help other stakeholders. One of these individuals is Niklas Zennström, who was "groomed" in the Kinnevik sphere before co-founding the multinational corporation Skype, and is consistently giving back to the innovation system in Stockholm both as an entrepreneurial role model and as an investor through his venture capital firm, Atomico. Niklas Adalberth, the co-founder of Klarna, a successful financial technology company – one of the Stockholm "unicorns," founded Norrsken Foundation as a way to give back to new entrepreneurs wanting to create social impact. Norrsken Foundation is a fund focusing on investments in so-called impact entrepreneurs, or entrepreneurs building ventures with a social impact, and has a popular co-working space tied to it, the Norrsken House, located in central Stockholm. The development of these massive MNCs, the resulting entrepreneurial ecosystems and clusters, and powerful Swedish investors have all helped solidify Stockholm as one of the most successful and innovative business systems in the world (Wall, 2017).

Relationships between spokes

From this framework, the success of Stockholm as an internationally significant innovation hub for sustainability can be observed in the close collaboration among research centers, private firms, and national and city governments. In order to have made these rapid gains as a capital city and to have become such a prolific hub for sustainability and innovation, Stockholm's hub elements have historically aligned and complemented each other to drive forward sustainability and innovation. From a government perspective, the policies put in place to drive technological and telecommunication advances have fostered technically mature consumers who are naturally desirous of new tech products. This consumer-driven demand has given Stockholm the opportunity to serve as a test market for new tech products. These "test beds" are areas dedicated to trying out innovative technology in the real world. The ICT Test Bed in Kista is a prime example of the confluence of the private sector, government, and academia. The initial birth of this cluster was formed with the establishment of Ericsson, which with its success attracted more IT and telecom companies to the area, which in turn eventually attracted academic facilities like KTH and Stockholm University to establish local campuses focused on ICT. The ICT Test Bed is – to a large extent – sponsored and coordinated by the local government, and consists of over 40 partners and 60 projects, all with the goal of developing tomorrow's sustainable and smart cities (Urban ICT Arena, 2016a).

International collaboration is a central tenet of the national innovation strategy for Sweden (Regeringen, 2004), and "few countries are in as good a position as Sweden to benefit from the ongoing process of internationalization and the emergence of the knowledge-based economy." While perpetual competition exists and innovation and growth are needed to outpace that competition, Sweden's "tradition of openness and . . . broad knowledge base has given us a head start over many other countries" (Regeringen, 2004). In a country with only 10 million inhabitants, Sweden has produced many large and reputable international companies. A strong export-oriented mindset is prominent in most new ventures from Stockholm and Sweden, says Johannes Schildt, CEO and co-founder of KRY, "because we live in a small country, you have to go to other countries in order to build a big company, so many people tend to think globally from the start," (Index Ventures, 2016). This sentiment is echoed by Niklas Zennström as he notes that "in the UK, you can still build a significant business by just being in the UK, but in Sweden you have to go abroad – and I think that's a benefit" (Index Ventures, 2016). This same export mindset is reflected in the Swedish public sector. Business Sweden, a merger of the Swedish Trade Council and Invest Sweden, is commissioned to support companies in reaching export markets and growing internationally and has a physical presence in roughly 50 world markets (Business Sweden, 2018). The Swedish government, through the Swedish Energy Agency (Energimyndigheten), has recently launched cleantech hubs in select major international cities, including San Francisco and Shanghai. Similarly, Vinnova has an international presence supporting Swedish startup companies looking to export to larger markets through the Nordic Innovation House, which is a soft-landing spot for startups in Silicon Valley, New York City, and recently launched locations in Singapore and Hong Kong.

Innovation and technology test beds

Various components of the hub are required to collaborate in order to develop connectivity and ICT. As Stockholm looks to support sustainable urban development, concepts like the Urban ICT Arena take root and flourish. This organization was founded and funded through a joint collaboration of academia (KTH and Stockholm University, RISE), businesses (ABB, Ericsson, IBM Sweden), and government (Stockholm City, The County Administrative Board – Länsstyrelsen Stockholm) (Urban ICT Arena, 2016a). As a collaborative organization, it has three components to its mission: develop sustainable cities, boost innovation, and secure future jobs (Urban ICT Arena, 2016b). As a part of the project, the founding partners contribute with various test beds in Kista, where new technology and innovations can be tested in real life. The 5G Proof-of-Concept Network by Ericsson is one of these test beds, where new innovations that require a 5G (or industrial internet) infrastructure can be tested, like the Not Boring 5G Bike, set to be the world's first 5G-connected bike. By leveraging the Ericsson 5G test bed to experiment with the bike's capabilities, the companies involved can develop various additional applications for smart and sustainable urban transportation.

The Urban ICT Arena is also taking part in a "Smart City Analytics and Insights" project, developing "innovative digital and mobile services based on the new open real-time city data available" (Urban ICT Arena, 2016c). In the project, a number of strategically located vehicle traffic flow measurements and WiFi-based sensors have been placed, and traffic data is collected in real time. This data, supported and analyzed by the IBM Bluemix platform, can be further analyzed and "turned into practical use cases reducing transport emissions and increasing the quality of life for citizens" (Urban ICT Arena, 2016d).

These efforts lead directly to the overlap with industry and academia in terms of developing a talent core. According to Wall (2017), access to competence is central for creating and sustaining a successful innovation hub. Fölster et al. (2017) further confirms this notion. This insight comes as no surprise, as Stockholm is the host of several prominent institutions for higher learning. One could argue that technology competence is the core of creating an innovation hub, and while this might be true, it is essential to have technological competence accompanied by complementary competence in areas such as business, natural sciences, and the social sciences, as well as rich design and music competence. Hedberg (2018) is of the opinion that the combination of competences that Stockholm possesses is essential for driving innovation. Equally important is the attractiveness of the city in terms of livability; "if the inhabitants enjoy a high quality of life then the city is more likely to attract and keep talent" (Hedberg, 2018). Stockholm has been highlighted as one of Europe's most beautiful cities regarding its landscape and natural waterways, architecture, and well-developed logistics for pedestrians and commuters. Stockholm fosters a walkable city layout, with an adopted strategic plan "Promenadstaden Stockholm" which is geared towards sustainability. This emphasis on walkability contributes to innovation, as various innovation hubs and actors in the city benefit from proximity to one another.

The effects of the implementation of multi-faceted policies in a supportive cultural system can be seen in the radical successes in Stockholm of innovators and entrepreneurs. The business landscape of Stockholm has even been transformed because of the confluence of academic, industry, and governmental decisions grafted onto the baseline societal cognizance of the environment. Specific to this is the generation of clusters from which innovation can be sparked and fostered.

Limitations

While Stockholm is demonstrably one of the most forward-thinking and successful capital cities for sustainability innovation, it is not immune to challenges. Communication between startups and large corporations is stove-piped, and the actors have historically not interacted as effectively as what would be typical in a more open environment, such as Silicon Valley, where large companies routinely partner with or acquire startups, fostering sustainable and continuous growth. According to Wall (2017), while large corporations in Stockholm have R&D departments and engage in university cooperation, they tend to be more limited in terms of their purchasing power for innovations and are not as open to acquire external innovations, which is frequently the case in other innovation regions such as Silicon Valley. Smaller startups have had difficulties reaching large corporations or governmental institutions as customers due to public procurement guidelines, which bottlenecks potential growth for some startups. Still, there are several initiatives that have been put in place to bridge the gaps between large corporations and startups. One such initiative is "Ignite Sweden" which now matches large corporations with startups to facilitate awareness, investment, and growth – funded by Vinnova (Vinnova, 2018b).

From a governmental perspective, it can be argued that the "safety net" of government programs allows citizens and companies to be less risk averse in terms of entrepreneurship. This government stability and benefits package, however, entails high taxes on small business and limits the ability of firms to solicit the highest-quality talent. In an effort to bring attention to this problem and provide a potential solution, the founders of Spotify held a campaign in Stockholm in 2016 to encourage the government to lower taxes for stock option programs for startups. Because startups typically do not have the ability to offer high salaries to high-quality talent, offering stock options provides potential talent an added incentive to join a startup (Ewig et al., 2016).

The lack of diversity among industries in Stockholm's corporate success stories presents another challenge. Unicorns have primarily emerged from software innovations and not from technologies that primarily support sustainability. However, the innovation-friendly climate also promotes cleantech companies, as evidenced by the many industry-agnostic success factors that have allowed the rapid rise of technology startups in Stockholm. One key indicator is investment momentum: According to research by The Nordic Web, Stockholm Business Region (which is wholly owned by the city of Stockholm), and SUP46, total investment in Stockholm's startups for the first six months of 2016 reached $1.2 billion, eclipsing the total amount raised throughout 2015 ($892 million). In fact, investment has quadrupled since 2014, which saw a comparatively meager $311 million invested in the city's startups. "I think it fair to say that Stockholm and Sweden have elevated themselves above everywhere else in the Nordics now and definitely should be considered on a level with London and Berlin as one of Europe's main ecosystems," said Neil Murray, The Nordic Web's founder (Index Ventures, 2016).

Conclusion

As evidenced in this chapter, the synergies among government policies (local and national), private sector initiatives, and academic research and programming have propelled Stockholm into its current standing as a hotbed for sustainability innovation. In particular, as the capital city of Sweden, Stockholm has benefitted from its proximity to key institutions driving innovation and sustainable practices. Its geographic location and position as capital city has allowed it to benefit from the early construction of public transportation infrastructure, access to private capital through commercial banks, and local inventions that eventually propelled high-tech startups to flourish within clusters formed around key multinational corporations. Furthermore, policies and targets set by the national government inspired Stockholm's government to match its own policies to those ambitious levels, or even to exceed them. Important distinctions, such as the Green Capital Award in 2010, set the bar high for Stockholm to push the boundaries for sustainability innovation even further with the target to become a fossil fuel free city by 2040. Within Stockholm's aspirational environment, ambitious individuals and firms have capitalized on opportunities to combine expertise in computer programming, engineering, and various technologies with a focus on environmental sustainability and a concern for the social implications of their business ventures in an effort to build commercially viable ventures, some of which have developed into unicorns. In the current phase of maturity, governmental, academic, and corporate actors in Stockholm are now channeling know-how and resources into creating an infrastructure to support new ventures that have the potential to drive innovation to the next level, while integrating societal and environmental goals into their business models. It is evident that Stockholm is an example of a capital city in which the interactions of the spokes of the hub have been mutually reinforcing and therefore resulted in a dynamic, entrepreneurial, and environmentally sustainable culture.

References

Almi. (2018). *Almi invest raises SEK 650 million for a GreenTech venture capital fund*. Retrieved November 2018 from www.almi.se/en/almi-invest/news/almi-nvest-raises-sek-650-million-for-a-greentech-venture-capital-fund/.

Andersson, M. et al. (2015). *Sweden: Decoupling GDP growth from CO2 emissions is possible*. Retrieved November 2018 from http://blogs.worldbank.org/climatechange/node/1005.

Atlantic Council. (2018). *The global innovation sweepstakes*. Retrieved October 2018 from www.atlanticcouncil.org/images/publications/The-Global-Innovation-Sweepstakes.pdf.

Atomico. (2017). *Billion dollar software companies*. Retrieved December 2017 from www.atomico.com/explore-d3.

Business Sweden. (2018). *Grow globally*. Retrieved November 2018 from http://beta.business-sweden.se/grow-globally/.

Cohan, P. (2017). KTH, the MIT of Stockholm, produced these three mind-blowing startups. *Inc. Magazine*. Retrieved March 2018 from www.inc.com/peter-cohan/kth-mit-of-stockholm-produced-these-three-mind-blowing-startups.html.

Davidson, L. (2015, June 28). *How Sweden became the startup capital of Europe*. Retrieved November 2017 from www.telegraph.co.uk/finance/newsbysector/mediatechnologyandtelecoms/11689464/How-Sweden-became-the-startup-capital-of-Europe.html.

Ek. (2018, May 10). Interview with Monica Ek, Professor of Wood Chemistry and Head of the Department of Wood Chemistry and Pulp Technology at the KTH Royal Institute of Technology.

Ericsson, K. (2018, May 10). Interview with Lisa Ericsson, Head of Department at KTH Innovation, KTH Royal Institute of Technology.

Ericsson, K. et al. (2016). The introduction and expansion of biomass use in Swedish district heating systems. *Biomass and Bioenergy* 94, 57–65. Retrieved November 2017 from www.sciencedirect.com/science/article/pii/S0961953416302793.

European Commission. (2018). *European Green Capital – Winning Cities*. Retrieved August 15, 2018 from http://ec.europa.eu/environment/europeangreencapital/winning-cities/.

Ewig, A. et al. (2016). *Options pay laws attacked in Sweden as startups call for change*. Bloomberg. Retrieved October 2018 from www.bloomberg.com/news/articles/2016-05-11/startups-urge-sweden-to-cut-stock-option-taxes-or-face-exodus.

Felten, B. (2015). *Stokab helps build a smarter Stockholm*. Retrieved October 2018 from www.stokab.se/Documents/Nyheter%20bilagor/Stokab%20helps%20build%20a%20smarter%20Stockholm.pdf.

Fölster, S. et al. (2017). *The geography of Europe's brain business jobs*. Stockholm: European Centre for Policy Reform and Entrepreneurship. Retrieved November 2017 from www.ecepr.org/wp-content/uploads/2017/11/Brain_business_jobs_final.pdf.

Government Offices of Sweden. (2018). *Sweden's carbon tax*. Retrieved October 2018 from www.government.se/government-policy/taxes-and-tariffs/swedens-carbon-tax/.

Gullberg, A. (2017, November 2017). Interview with Anders Gullberg, MISTRA Sams Project.

Hedberg, P. (2018, May 2018). Interview with Pär Hedberg, CEO at STING – Stockholm Innovation and Growth.

Hofstede Insights (2018). *Compare Countries (on-line database)*. https://www.hofstede-insights.com/product/compare-countries/.

Hugosson, B. (2018, January 2018). Interview with Björn Hugosson, Head of Climate at Stockholm City.

Index Ventures. (2016). *How the Swedish capital became Europe's unicorn powerhouse*. Retrieved November 2018 from www.indexventures.com/index-insight/stockholm-syndrome.

Lee, A. (2013). *Welcome to the unicorn club – learning from billion-dollar startups*. TechCrunch. Retrieved October 2018 from https://techcrunch.com/2013/11/02/welcome-to-the-unicorn-club/.

Milne, R. (2017). The Wallenbergs: Where money meets Swedish science. *Financial Times*. Retrieved April 2018 from www.wallenberg.com/sites/default/files/files/pdf/ftrp-4553finallocked.pdf.

Metzger, J. et al. eds. (2013). *Sustainable Stockholm: Exploring urban sustainable development in Europe's greenest city*. New York: Routledge.

OECD. (2013). *Green Growth in Stockholm, Sweden, OECD Green Growth Studies*. Paris: OECD Publishing. http://dx.doi.org/10.1787/9789264195158-en.

Regeringen. (2004). *Innovative Sweden – a strategy of growth through renewal. Ds 2004:36*. Stockholm: The Ministry of Industry, Employment and Communications and the Ministry of Education. Retrieved March 2018 from www.regeringen.se/49bbba/contentassets/d0f7ce031b504663893e4ff83f89cd93/innovative-sweden – a-strategy-through-renewal.

Regeringen. (2016). *Pressmeddelande: Staten och kapitalet, reformer för innovation och hållbar tillväxt*. Stockholm: The Ministry of Industry, Employment and Communications and the Ministry of Education. Retrieved March 2018 from www.regeringen.se/pressmeddelanden/2016/03/staten-och-kapitalet–reformer-for-innovation-och-hallbar-tillvaxt/.

Regeringen. (2018a). *Regeringen – om myndigheter med flera – industrifonden*. Stockholm: The Ministry of Industry, Employment and Communications and the Ministry of Education. Retrieved March 2018 from www.regeringen.se/myndigheter-med-flera/industrifonden/.

Regeringen. (2018b). *Finansiering av forskning i Sverige*. Stockholm: The Ministry of Industry, Employment and Communications and the Ministry of Education. Retrieved April 2018 from www.regeringen.se/regeringens-politik/hogskola-och-forskning/finansiering-av-forskning-i-sverige/.

Robecosam. (2015). *Measuring country intangibles – Robecosam's country sustainability ranking*. Retrieved March 20, 2018 from www.robecosam.com/images/Country-Sustainability-Paper-en.pdf.

Saminvest. (2018). *Saminvest – om Saminvest*. Retrieved March 2018 from http://saminvest.se/sv/om-saminvest.

SCB. (2017). *Folkmängd i riket, län och kommuner 31 december 2017 och befolkningsförändringar 1 oktober – 31 december 2017*. Retrieved March 2018 from www.scb.se/hitta-statistik/statistik-efter-amne/befolkning/befolkningens-sammansattning/befolkningsstatistik/pong/tabell-och-diagram/kvartals–och-halvarsstatistik–kommun-lan-och-riket/kvartal-4–2017/.

SEB. (2018). *Our history*. Retrieved October 2018 from https://sebgroup.com/about-seb/who-we-are/our-history.

Svebio. (2018). *Bioheat – about*. Retrieved March 2018 from www.svebio.se/om-bioenergi/biovarme/.

Sweden.se. (2018a). *Sweden's official website*. Retrieved October 28, 2018 from https://sweden.se/society/energy-use-in-sweden/.

Sweden.se. (2018b). *Sweden's official website*. Retrieved March 20th, 2018 from https://sweden.se/nature/sweden-tackles-climate-change/.

United Nations. (2018). *Sustainable development milestones*. Retrieved January 2018 from https://sustainabledevelopment.un.org/milestones/humanenvironment.

Urban ICT Arena. (2016a). *Founding partners*. Retrieved January 8, 2019 from http://www.urbanictarena.se/category/partners/founding-partners/.

Urban ICT Arena. (2016b). *Our three aims*. Retrieved January 8, 2019 from http://www.urbanictarena.se/our-three-aims/.

Urban ICT Arena. (2016c). *5G POC*. Retrieved January 8, 2019 from http://www.urbanictarena.se/5g-poc/.

Urban ICT Arena. (2016d). *Grow smarter – Sustainable and smart solutions for an urban environment*. Retrieved January 8, 2019 from http://www.urbanictarena.se/grow-smarter-comes-urban-ict-arena/

Vinnova. (2018a). *Our role*. Retrieved November 2018 from www.vinnova.se/en/about-us/swedens-innovation-agency/the-role-of-vinnova/.

Vinnova. (2018b). *Ignite Sweden*. Retrieved October 2018 from www.vinnova.se/p/ignite sweden/.

Visit Stockholm. (2018). *The art of the subway. Discover Stockholm's metro – the world's longest art exhibition*. Retrieved October 2018 from www.visitstockholm.com/art-in-the-subway/.

Wall. (2017, November 17). Interview with Marie Wall, Deputy Director – Startups, Swedish Ministry of Enterprise.

World Bank. (2015). *Research and development expenditure (% of GDP)*. World Bank Data. Retrieved November 2018 from https://data.worldbank.org/indicator/GB.XPD.RSDV.GD.ZS.

6 Using public procurement to promote sustainability in Copenhagen

Marta Andhov and Alexandra Andhov

Capital cities enjoy a special place as a hub, where governments, private corporations, NGOs, media and other entities are present and interact.[1] According to the theory presented in Chapter 1 of this volume, capital city hubs theoretically represent an ideal space for innovative thinking and developing sustainable policies and practices. In this chapter, we assess Denmark's capital Copenhagen and question whether it utilizes its hub in pursuance of sustainability goals. Ultimately, we conclude that the Danish capital successfully deploys innovative public procurement practices to integrate different parts of the hub as a way of implementing effective sustainability policies.

We first introduce Copenhagen and its urban and political character. Next, we explain how procurement policy can be a tool for supporting and improving sustainability practices. Public procurement, in simple terms, represents state procurement of goods, services and construction on behalf of a public authority, body or agency. We argue that public procurement regulation can be used to nudge diverse parties to support and advance sustainability goals. Public procurement – and possibly private procurement, as well – can be used as a thread linking all the diverse constituencies of the hub described in Chapter 1 to develop sustainability solutions and create an environment whereby sustainability represents an important and relevant goal.

In short, this chapter aims to show how public procurement can be used as a tool to deliver sustainability. To develop our argument, we introduce the existing public procurement regulation and explain how it has changed the existing procurement policies in Copenhagen and the relationship between the city and its suppliers, subsequently triggering other changes with a direct impact on sustainability. We show how public procurement can be understood as a regulatory rationale for sustainability in capital cities, where the majority of public authorities and their suppliers are concentrated. In making our argument, we analyze the behavior of other hub entities, such as universities, the media, NGOs and the international community, showing how they contribute to an ecosystem for enforcing public procurement policy. Before concluding, we analyze the existing regulatory obstacles that limit the use of public procurement in an even broader advancement of sustainability.

Copenhagen: a small innovative European city

In 2009, the Copenhagen metropolitan area (*Hovedstadsområdet*) was inhabited by close to 2.4 million people, which represents the largest metropolitan area of any Scandinavian city, according to the OECD (2009). The Copenhagen metropolitan region accounts for 44 percent of the Danish population, in an area that includes the cities of Copenhagen and Frederiksberg, as well as five adjacent former counties. Among 78 OECD metropolitan areas with populations of more than 1.5 million inhabitants, the Copenhagen metropolitan area ranks fourth in terms of its share of national output. The Copenhagen metropolitan area concentrates 80 percent of Denmark's high-tech firms and 70 percent of its private research and development, and is home to the best universities in the country.

Copenhagen City (*Københavns Kommune*) represents the largest area within the Copenhagen metropolitan area and has a population of 603,000 people within 86 km^2. Copenhagen City itself represents an administrative area, with the seat of government in the Copenhagen City Hall. The City Council is the highest political authority of Copenhagen City, with 55 members elected for four-year terms. The head of the City Council, the Lord Mayor, summons the meetings, organizes the discussions and determines the agenda at the City Council meetings. Since 2010, Frank Jensen has served as Lord Mayor of Copenhagen.

Copenhagen benefits from a diversified metropolitan economy with numerous multinational corporations and research institutions, as well as international and regional organizations. Copenhagen excels in the service sector, making it comparable to London or Paris. Business services remain the largest employer, whereas Copenhagen represents one of the European leaders in the area of biotechnology (OECD, 2009, 3). Copenhagen itself is a vivid and diverse city with an active cultural life.

Nevertheless, Copenhagen faces several challenges. Owner-occupied flat and single-family house prices in Copenhagen have been rising by around 10 percent a year since 2012 (OECD, 2016), which represents a challenge for the financial sustainability of Copenhagen's inhabitants. Furthermore, despite good environmental policies and many positive trends, such as success in decoupling emissions and farm inputs from economic activity, the pressures on the environment from manufacturing, agriculture, transport and energy persist. Furthermore, in a country with low car ownership and a high cycling rate, around one-fifth of the population is exposed to poor air quality (OECD, 2016, 34); levels of PM_{10} and NO_2 exceeded the EU limit values on a number of occasions, and in 2011 air pollution caused an estimated 3,200 premature deaths (WHO, 2013). Copenhagen is working to improve these figures. Additionally, in 2009, the City Council unanimously adopted the Climate Plan for Copenhagen, setting goals for achieving a 20 percent reduction in CO_2 emissions by 2015 and complete carbon neutrality by 2025 (Copenhagen Council, 2009). The first goal was successfully achieved. Copenhagen City is also a member of the C40, the world's leading climate network for megacities, hosting the C40's permanent office, and serves as a center for the network's global Business, Economy and Innovation Programme. The office works to foster cooperation between cities and the private sector.

Public procurement as a tool for delivering sustainability

In this section, we lay out our argument that public procurement works to bring together the various parts of the hub in pursuit of capital city sustainability policy. Public procurement refers to the process by which public authorities – such as capital cities, municipalities and government departments – purchase goods or services from companies. Examples include state building projects, purchasing furniture for the public prosecutor's office or contracting cleaning services for a public university.[2] Historically, there have been frequent attempts to link public procurement with government policy in areas as diverse as national industrial policy, reducing unemployment, support for small businesses, local development, employment of disabled workers, or equal pay for men and women as well as sustainability (McCrudden, 2007).

Denmark, a small welfare state with a population of 5.7 million, spends a substantial amount on its public procurement – approximately 290 billion DKK per year (~€39 billion) – and is viewed as a leader in integrating environmental technologies and policies into its procurements.[3] Public opinion in the country strongly supports environmental responsibility, as well as efforts to promote social responsibility and minimize corruption. By public support, we mean not only political backing, but also daily consideration of these issues among the media and the general population.

Copenhagen City has identified a clear set of sustainability goals, which are laid out in its publication *Solutions for Sustainable Cities in 2014* (3rd edition). Top priorities include providing integrated transportation solutions in the city and within its surrounding towns and areas, with a special emphasis on cycling. Due to the location of Copenhagen, the city also made water quality one of its priorities. Other goals highlight promoting renewable energy, limiting waste and securing the longevity of local construction projects. Additionally, Copenhagen has declared a target of becoming completely carbon neutral by 2025.[4]

Denmark has a well-functioning procurement system that is advanced in its strategic and sustainability dimensions.[5] Certain parts of public procurement are conducted at the local level, as numerous administrative obligations are decentralized. Denmark's 98 municipalities are responsible for the delivery of services, such as road maintenance, cleaning and social services (childcare, primary school, care of elderly and specialized treatment of the disabled). At the same time, a part of procurement remains centralized by the state in the capital city of Copenhagen. Central purchasing bodies amongst others include the largest Danish public purchaser the State and Municipal Purchasing Service (SKI – *Statens og Kommunernes Indkøbsservice A/S*).[6] The Danish Competition and Consumer Authority (*Konkurrence og Forbrugerstyrelsen*) plays a crucial role as it is responsible for oversight, guidance and support in public procurement in general.[7] At the ministry level, the Agency for Modernization (*Statens Indkøb*) is in charge of procurement policy and of the aggregation of procurement needs for government bodies.[8] Finally, the Danish Public Procurement Complaints Board (*Klagenævnet for Udbud*) is an independent, administrative, quasi-judicial complaint body that deals with complaints and challenges regarding non-compliance with procurement law.

Copenhagen as a capital city actively applies public procurement as a strategic tool for delivering on its sustainability agenda. Such sustainable public procurement is a process by which organizations meet their needs in a cost-effective manner, while also striving toward environmental and social objectives, including reduction of greenhouse gas emissions, energy and resource efficiency, job creation, social inclusion and equality (Andrecka, 2017; Andrecka & Mitkidis, 2017).

In Copenhagen, the 55 members of Copenhagen's City Council hold 79 positions/seats in the seven committees (called administrations). The seven committees are:

- The Finance Administration
- The Culture and Leisure Administration
- The Children and Youth Administration
- The Health and Care Administration
- The Social Services Administration
- The Technical and Environmental Administration
- The Employment and Integration Administration

Each committee has a specific area of responsibility; however, the Finance Administration represents an overarching committee. The Finance Administration and the Sector Skills Council provide overall political management, the central administrations execute their decisions, whereas the managers of the decentralized administrations and institutions share day-to-day purchasing responsibility with the local day-to-day unit operators.[9] Public procurement in Copenhagen is spread out over seven administrations, 10 procurements units, approximately 60 contract lawyers and thousands of decentralized procurers (Figure 6.1). Furthermore, when contracting, Copenhagen uses a contract appendix, which obliges a supplier to implement specific principles of corporate social responsibility.[10]

Beyond the City Council, corporations, various non-profit groups and the media play a role in the procurement process. Whether advocating for sustainability goals or simply trying to sell green products to the city government, these groups have a strong impact on the outcome of city purchasing. Corporations seek to sell their products to the city. While some of them come up with innovative sustainability ideas on their own, others are inspired by city sustainability regulations to meet the targets set by public officials. Non-profit groups, such as DanWatch, investigate how well the city and the corporations it hires are actually living up to their published human rights and environmental goals. The media, in turn, works with the NGOs to inform the broader public about what the city and corporations are doing. The media rely heavily on the expertise developed

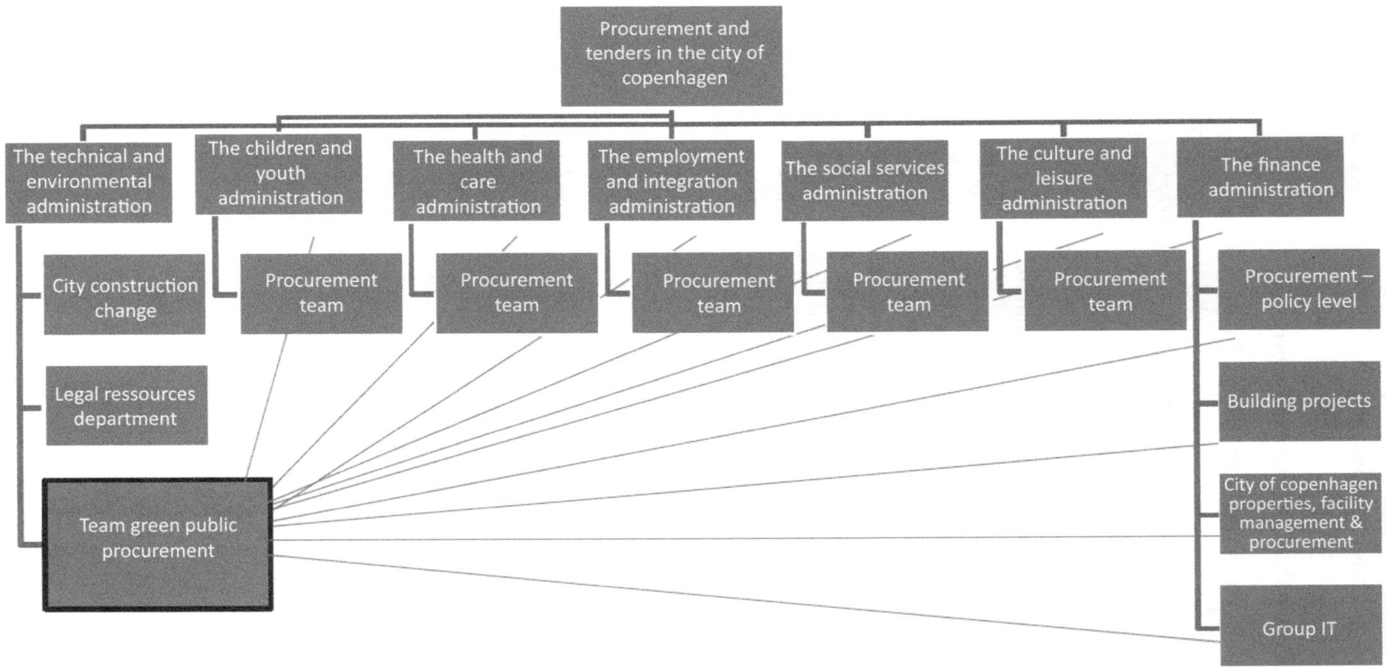

Figure 6.1 Organization of public procurement in Copenhagen city

Source: http://www.procuraplus.org/fileadmin/user_upload/Activities_files/Events/Rome_2016/Procura__2016_Key_Note_Speech_Copenhagen.pdf

by DanWatch and local universities to trace the intricacies of the public procurement process and bring to public attention important issues when they arise.

Copenhagen's public procurement process

Since Copenhagen spends more than €1.5 billion annually, public procurement is a strategic tool for achieving sustainability, and therefore a prime topic of public discussion.[11] In the following subsections, we first analyze Copenhagen's procurement policy for sustainability. In the next section, we assess the city's corporate social responsibility (CSR) contract appendix.

Copenhagen's procurement policy for sustainability

Environmental considerations began to influence the work of state procurement officers in Denmark in 1991. Denmark's first Green Public Procurement National Action Plan (GPP NAP) was adopted in 1994. Between 1997 and 2005, GPP guidelines (*Miljøvejledninger*) were developed for 46 product groups. In 2008, as a part of the EU Sustainable Consumption and Production Action Plan, a target of 50 percent for green procurement was defined and applied.[12] Following these advances, the Danish Ministry of Finance put forward a Strategy on Intelligent Public Procurement in October 2013, stating that all public procurement should meet three overall objectives:

- Efficiency
- Innovation and quality development
- Sustainability[13]

Thus, since 2013, sustainability has been strongly promoted anytime a public body decides to purchase goods or services. The Ministry of Environment and Food (MEF) has established a Partnership for Public Green Procurement, including 12 municipalities, two regions and a water supply company.[14] The partnership accounts for 17 percent of all public procurement in Denmark, or 50 billion DKK (~€6.7 billion) in total.[15] Copenhagen is a member of this partnership and has committed itself to follow the jointly agreed environmental public procurement criteria within the categories listed in Table 6.1.

Copenhagen plans to use sustainable public procurement as a tool to attain its five major goals for 2025:

- Reduce energy consumption in city administration buildings by 40 percent compared to 2010
- New municipal buildings erected up to 2015 must meet the 2015 building code requirements, and new buildings started between 2015–2020 must meet the 2020 standards
- All city administration vehicles must run on electricity, hydrogen or biofuels
- Energy consumption for street lighting in Copenhagen should be 50 percent of 2010 levels
- Municipal buildings must install a total of 60,000 m^2 of solar cell panels[16]

Table 6.1 Public procurement goals for waste prevention, reuse and recycling

Targeted area	*Goals for waste prevention, reuse and recycling*
Food	All primary and transport packaging should be free of PVC, phthalates and other items on the list of undesirable substances (cf. Danish EPA).
Transport	Members should set criteria for the disposal of old vehicles.
Construction	Members must set criteria for a number of areas, including waste and recyclability, and they may set the criterion that building materials shall not contain PVC.
Products for children	All toys should be without phthalates, brominated flame-retardants, heavy metals hazardous to health, organic solvents or azo colors. Diapers must comply with the criteria of the Nordic eco-label "Svanen."
Cleaning products and services	Cleaning products must comply with the criteria of the Nordic eco-label "Svanen" or the EU Eco-label "The Flower."

Source: http://ec.europa.eu/environment/life/project/Projects/index.cfm?fuseaction=home.showFile&rep=file&fil=PLASTIC_ZERO_analyzed_in_denmark_case_copenhagen_final_.pdf

Therefore, the re-evaluation of its procurement processes is listed as a key milestone goal in the city's 2025 Climate Plan. The Plan also states a desire to re-evaluate the city's tendering processes with methods including Life-Cycle Costing (LCC) and functional tendering (Chrintz, 2011). LCC is an integrated framework of concepts and techniques to address environmental, economic, technological and social aspects of products, services and organizations.[17] This concept directly contributes to achieving sustainability in the city, but also reshapes the approach of the businesses as they seek to increase their own sustainability and increase their competitiveness in applying to provide goods and services to the city.

Labor clauses are a particularly hot topic in Copenhagen, and in the rest of Denmark. In fact, both central government entities are subject to a "comply or explain" principle (*følg eller forklar*), which obliges these entities to either use social clauses in training and apprenticeships agreements in relevant procurement (e.g. construction projects, such as building hospitals, infrastructure, etc.) or explain why they do not use such clauses.[18] Other government authorities and government entities, including companies that are fully owned by state authorities and are not in competition, must apply labor clauses for contracts in the construction sector.[19] For municipal and regional contracting authorities there is no such obligation to apply labor clauses. However, it is recommended that all municipalities and regions apply them in building and construction contracts.[20]

City contract appendix: the supplier's corporate social responsibility

In addition to EU, Danish and city-level sustainable policies, Copenhagen has been using as an additional legal tool: the contract appendix, which requires suppliers' compliance with corporate social responsibility (CSR) goals.[21] The contract appendix requires all suppliers working with the city to comply with specific sustainability requirements, which are based on the ten principles of the

UN Global Compact (UNGC),[22] the UN Guiding Principles for Business and Human Rights[23] and ILO conventions. The municipality of Copenhagen,[24] as well as SKI[25] and the Central Procurement Programme,[26] impose substantive supplier requirements regarding:

- Human rights (e.g. "the Supplier undertakes, in executing the contract, to respect basic human rights as per principles 1 and 2 of the UN Global Compact and as manifest inter alia [i.a.] in ILO Convention No. 100 [1951] on equal pay and No. 111 [1958] on discrimination")
- Labor rights including minimum wages (e.g. "the Supplier undertakes to respect basic worker rights, which means i.a. that the delivered works and parts thereof must be produced under conditions where the general principle of the right to a safe and healthy working environment is observed, cf. i.a. ILO Convention No. 155 [1981].")
- Environmental protection (e.g. "the Supplier must especially strive to prevent and combat pollution of air, water, soil and subsoil as well as vibration and noise nuisances")

If grounded suspicion arises that the requirements have not been met, the supplier must, upon written request by the City of Copenhagen, submit a written account of conduct, including documentation of:

- The processes and/or methods under which the goods, services and/or building and construction works in question have been produced
- The materials used in executing the contract

In this way, Copenhagen conducts oversight of the actions taken by its suppliers. An important factor is also that the supplier must account for its sub-suppliers (the entire supply chain). The supplier must aim to fulfill the city's requirements, including through the selection of sub-suppliers or the selection of components for the delivered works. Furthermore, the supplier must account for certificates relating to production processes and methods as well as any choice of materials.

If a supplier is found in serious breach of its CSR requirements, the City of Copenhagen may terminate the contract in full or in part. The appendix also stipulates what actually formulates a serious breach, namely (i) supplier's failure to meet the CSR requirements and (ii) supplier's failure to meet the requirement to provide necessary documentation. However, it is important to note that given the language used in that provision, the city is not obliged to automatically terminate the contract, but is rather provided a choice. The City of Copenhagen carries out a specific assessment in each individual case, including reviewing whether it wishes to engage in dialogue with the supplier or automatically terminate the contract. If the city does not terminate the contract, it shall impose a fine. The fine is automatic and not discretionary. The fine is defined as a daily penalty until the supplier rectifies the breach. Each contract has different set amounts, depending on the nature and scope of the contract.

Additionally, the supplier's payment of a fine does not preclude the City of Copenhagen from claiming compensation from the supplier under the general provisions of Danish law. The city is entitled to withhold payment in order to honor valid claims from the supplier's or sub-suppliers' workers. Given these conditions, the standards to which the suppliers of the city are held can be considered high. It is important to note that such standardization of contractual relationships – through both procurement processes and use of contractual warranties – substantially contributes to the awareness of sustainability and CSR, as well as its enforcement.

Universities as part of the sustainability and public procurement discussion

Public procurement directly covers universities in Denmark. The majority of institutions of higher learning are public, and therefore, whenever they acquire any goods or services, they must adhere to the existing procedures and regulations. Universities may decide to buy directly or through the central purchasing bodies. The major universities in Copenhagen – the University of Copenhagen, Technical University of Denmark (DTU) and the Copenhagen Business School (CBS) – have established their own administrative centers that carry out procurement. In assessing their respective procurement policies, we discovered that CBS has not published explicit guidelines, whereas the University of Copenhagen made public its Purchasing Policy, which lays down the framework, norms and guidelines for all purchases.

The University of Copenhagen Purchasing Policy stresses the principle of "highest quality at lowest cost," which means that the following requirements must be fulfilled:

- Correct and suitable specifications and quality
- Lowest total cost from "cradle to grave," thereby taking the environment and sustainability into consideration
- Standardization
- Exposing the purchasing agreements to ongoing competition
- The greatest degree of independence from single suppliers[27]

From these provisions, one can see that other public institutions in addition to the city are directly proclaiming sustainability, energy and the environment as considerations in procurement decisions. Like the city, in 2008 the University of Copenhagen adopted a campus strategy entitled "Green Campus 2020," which defines numerous sustainability goals for the university, including specific targets for energy efficiency and CO_2 reductions.[28] As part of Green Campus 2020, the university collaborates with both internal and external partners, while supporting further research in innovative solutions. The University of Copenhagen is a member of International Alliance of Research Universities (IARU), which also has been focused on sustainable development.[29]

In addition to greening its operations, sustainability is integrated into research and teaching at both universities to provide a forum for academics, politicians and practitioners to discuss various aspects of sustainable procurement. In 2014, the University of Copenhagen organized a sustainability conference, together with IARU, focused on transforming universities around the world. In 2017, the Faculty of Law hosted a conference entitled "Procurement beyond price: sustainability and CSR in public purchasing." Both institutions provide an array of courses for students that cover supply chain management, public procurement and other relevant subjects. There are more than 60 courses at the university that directly cover sustainability, environment or natural resources. Moreover, numerous Danish foundations have supported sustainability research, including Velux Foundation, Carlsber Foundation and Dreyers Fund. The question of sustainability and innovation through procurement have become a part of the university environment.

Key obstacles in achieving Copenhagen's sustainability efforts

There are two key obstacles in the area of sustainable public procurement, which limit the way that Copenhagen can pursue a sustainability agenda through this mechanism. The first is the "link to subject matter" concept that is a major limiting principle for sustainability considerations in the EU public procurement directives. The second obstacle is enforceability issues related to the sustainability contract clause. Both are analyzed in the following subsections.

Link to subject-matter

Copenhagen's procurement must be in compliance with Danish public procurement law, which itself is regulated at the EU level. To create a level playing field for all businesses across Europe, the EU law sets out minimum harmonized public procurement rules. These EU Public Procurement Directives organize the way public authorities, such as the city of Copenhagen, and certain public utility operators purchase goods and services.[30] They are transposed into national legislation (the Danish Public Procurement Act) and apply to tenders whose monetary value exceeds:

- €5,225,000 for work contracts[31]
- €135,000 for central government contracts
- €207,000 for local and regional government contracts[32]
- €750,000 for social and other specific service contracts[33]

These national rules also have to respect the general principles of EU law such as transparency, non-discrimination, open competition and mutual recognition.

Since every year over 250,000 public authorities in the EU spend around 14 percent of GDP on the purchase of goods and services, European public authorities hold significant market power.[34] They have the potential to influence a shift in supply towards increasingly green solutions by demanding such products and services. Within Scandinavian markets, a large part of this purchasing power is at the municipal level, as individual municipalities have high procurement autonomy and control. Nevertheless, a certain degree of distrust still lingers concerning both the legitimacy of using public contracts to pursue sustainability-related objectives and its effectiveness.[35]

The Court of Justice of the European Union (CJEU) developed the "link to subject matter" concept in its case law regarding award criteria for public contracts. In the Court's first judgment in this area, the *Concordia* case, where the public entity used environmental considerations, namely the emissions of nitrogen oxide and noise pollution, among the criteria for the contract award, the CJEU established that:

> Where the contracting authority decides to award a contract . . . it may take criteria relating to the preservation of the environment into consideration, provided that they are linked to the subject-matter of the contract.[36]

At the same time, using these criteria may not grant the contracting authority unrestricted discretion in choice of a tender; the criteria must be explicitly mentioned in the contract notice or tender documents, and they must comply with the fundamental European Union Treaty principles, in particular non-discrimination.[37] Further case law development in this area included a challenge to the environmental awarding criteria, when the public entity had allocated 45 percent of the award for bidders' ability to produce renewable electricity in amounts, which exceeded the volume required under the contract (*EVN and Wienstrom* case).[38] The CJEU ruled that the focus on capacities of electricity which exceeded the public entity's requirements doomed the criterion to be *not* linked to the subject matter of the contract.

Critics argue that the "link to the subject matter" concept practically disables public procurement's effective pursuance of sustainability goals (Semple, 2016). They claim that this concept makes it impossible to include general CSR policies to the extent that they address matters beyond the specific needs of the public entity. However, what does "the specific needs of the public entity" mean in practice? In our understanding and in the light of the existing case law, the sustainability and/or CSR requirements must be connected to the contract awarded – meaning that the contract performance clause directly linked to the activities carried out under the contract, such as obtaining a recycling rate over a given percentage of materials disposed of during construction, powering a festival only with renewable energy or releasing in open source whatever intellectual property was developed under the contract, shall be permissible.[39]

It is worth noting that the understanding of the "link to subject matter" concept has expanded systematically from its establishment in the *Concordia* case. During the process of redrafting procurement directives, the CJEU dealt with the milestone *Dutch Coffee* case.[40] In this case, a public entity wanted to include award criteria for a supply contract of tea and coffee vending machines to relate to their organic and fair trade character. In its ruling, the CJEU confirmed that non-economic criteria which relate to a particular means of production (e.g. organic character) or distribution (e.g. fair trade labels) could be considered to be linked to the subject-matter of a contract.[41] In this judgment, the CJEU expanded the concept of link to subject matter of a contract by underlining that there was no requirement for award criteria to relate to a core characteristic of a product or something which alters its material substance. The newest developments in this area came from the EU's Directive 2014/24/EU, whereby among other things, a definition of the concept can be found which was influenced by the *Dutch Coffee* case. According to Article 67:

> Award criteria shall be considered to be linked to the subject-matter of the public contract where they relate to the works, supplies or services to be provided under that contract in any respect and at any stage of their life cycle, including factors involved in:
>
> a) the specific process of production, provision or trading of those works, supplies or services; or
> b) a specific process for another stage of their life cycle, even [when] such factors do not form part of their material substance.

It is important to read Article 67 in combination with Recital 97 of Directive 2014/24/EU, which includes certain limitations:

> [T]he condition of a link with the subject-matter of the contract excludes criteria and conditions relating to general corporate policy, which cannot be considered as a factor characterizing the specific process of production or provision of the purchased works, supplies or services. Contracting authorities should hence not be allowed to require tenderers to have a certain corporate social or environmental responsibility policy in place.

Consequently, the combined Article 67 and Recital 97 emphasize that public procurement contracts must relate to the goods, services or works that are being purchased, and cannot concern matters which fall outside of the scope of the procurement relationship and the public contract itself. Hence, even though the EU public procurement rules limit the scope in which sustainability can be introduced as a requirement for a public contract, we observe a continuous adaptation of the subject-matter concept by cities like Copenhagen as they work to achieve their goals.

Hampered enforceability and the importance of close media attention

Copenhagen's active work to incorporate sustainability considerations and clauses into its public procurement practices deserve praise. If nothing more, this effort increases awareness and debate among Danish public procurers on how to prevent violations in their supply chains and how to proactively take a lead in their communities. However, challenges do occur in the application of sustainability annexes or clauses with references to such international documents as the UN Global Compact. Even though, as noted earlier, if a corporation violated the conditions of an annex, constituting a material breach, and possibly leading to the termination of a contract, the question is whether and how often these clauses are enforced. The answer is that they are not enforced very often. Usually, the violations are regulated in stages. First, the supplier would be asked for explanation, if it continued the violations the contracting authority may impose a penalty for the violations or ultimately decide to terminate the contract. It is debatable if there is typically an interest in terminating a contract on sustainability grounds. This leads to the conclusion that the provisions in these CSR annexes often end up being nothing more than a public statement, which is only used when headlines in the newspapers demand that the contracting authority take action. Even with this glare of publicity, practice shows that public authorities are hesitant to terminate the public contract.

The latter may be seen in the example of a contract for information technology hardware (laptops, desktop computers, etc.) concluded between SKI and Dell. SKI in its tender applied the CSR annex. Subsequently, in October 2013, the independent investigative media and research center DanWatch brought attention to the working conditions in four of Dell's factories in China.

DanWatch was established in 2007 as a Danish media and research center specialized in investigative journalism. DanWatch monitors how states and companies impact human rights, the environment, and labor issues. Five areas of particular concern for DanWatch are i) textiles and garments, ii) food, iii) transport and shipping, iv) technology and v) energy. DanWatch has published numerous relevant reports that won considerable public attention. Many of these investigations focus on third countries, but some touch upon practices in Denmark, Scandinavia or greater Europe. In 2008, DanWatch published a report, commissioned by DanChurchAid and the Roskilde Festival, which investigated illegal mining practices in the Democratic Republic of Congo. DanWatch named numerous mobile phone producers, including Nokia, Motorola and Samsung, that all buy cobalt from DR Congo and thus are likely to support illegal export and mining practices, which involve severe human rights violations. More recent reports uncovered that Carlsberg's brewer in Nepal contributed to soot pollution in water supplies affecting a village near its factory.[42] The water samples were tested by a professor of Environmental Chemistry at the University of Copenhagen, as well as by a professor and wastewater expert at the Technical University of Denmark (DTU). This report garnered considerable media attention. Generally, DanWatch is considered a reliable media source and a valuable contributor to public discussion and oversight. However, DanWatch has also received some negative reviews, including allegations that it relies on biased NGOs for its information and non-transparent sources of financing.[43] DanWatch is supported both by public foundations[44] and by private parties.[45]

In Dell's case, according to the DanWatch report, the employees worked up to 267 hours per month and their hourly wage was between 7 and 8 DKK (~€1). As these conditions violated the provisions in SKI's CSR annex, action was needed. However, the challenge was that the annex did not provide an indication of how to proceed in these situations. SKI decided to request from Dell – along with several other suppliers – a report stating the conditions in the relevant factories, as well as action plans for how the identified issues would be resolved.[46] When the reports were handed in, they were reviewed and followed for about one year. Ultimately, it was concluded that the supplier conducted a successful self-cleaning and therefore, no contractual breach was stipulated.

In practice, there may be several reasons for the lack of enforcement of contract termination clauses.[47] First, it seems that not only in the area of sustainability or CSR but in general in public procurement, contracting authorities are eager to avoid terminating the contract if it is not absolutely necessary. Usually, a contracting authority will attempt to escape wasting time and resources in retendering the contract and rather will attempt to force the supplier to meet its contractual terms. Second, violations of the sustainability clauses, even if the annexes treat them as a material breach, are still perceived by the contracting authority as secondary elements to the contract delivery. This may lead to contracting authorities not being motivated to enforce them. If the latter is the case, it is the mindset and core values of the staff working with these issues that have to be changed, rather than the legal basis. Finally, even when there is a will from the contracting authority to take the sustainability clauses seriously, there may be a lack of opportunity or other substantial challenges to do so. For example, there may be no alternative supplier for the delivery of the public contract, or there may be a lack of knowledge or resources and/or tools to monitor and enforce compliance with the sustainability clauses.[48]

Besides DanWatch, there are numerous media outlets that follow both sustainability issues and public bodies' activities on a daily basis. Copenhagen, as well as the Danish media, have devoted a lot of time to building and educating the public on the topic of sustainability. In such print media as *Morgenavisen Jyllands-Posten*, *Politiken* and *Borsen*, there are daily stories emphasizing the necessity for sustainable operations across all the sectors. In Copenhagen, universities together with NGOs and media organize conferences and platforms for discussing sustainability, which are attended by small and large corporations. All of these efforts create public demand for more sustainability policies.

Conclusion

In this chapter, we analyzed the use of public procurement as a legal tool to achieve sustainability in Copenhagen. We argue that public procurement is a thread connecting various entities active in the city to promote pro-sustainability policies and behavior. Copenhagen is a good example of the capital city hub, as it is the physical place where a number of entities have the ability to discuss and adopt sustainability policies through the public procurement process.

As a city, Copenhagen has been active and vocal about its approach to sustainability. It uses legal tools – procurement and contractual provisions – to implement sustainability policies in relation to private and public partners. These partners are required to promote sustainability when carrying out contractual obligations for the city. The procurement and contractual provisions provide for greater awareness and compliance among private parties. Moreover, media and research organizations based in the city use their proximity and direct links to the main public agencies and corporations to conduct oversight and reflect on the existing policies and practices. When cases of non-compliance are found and brought to public attention, the interconnection among different partners active in the hub and involvement of the public often helps to ensure that Copenhagen meets its sustainability goals. In this sense, the hub creates a virtuous circle for sustainability. Ultimately, the adoption of some sustainable procurement policies leads to a cascade of more policies as the first ones start to take effect. Similarly, exposing circumventions and discussing them in the media raises awareness about compliance issues and directs more public effort to bringing corporations back into line with their public procurement commitments.

After providing an overview of the existing sustainability policies in Copenhagen, we focused on explaining the scope of public procurement in the capital city and explained how different public bodies incorporate sustainability considerations into their procurement policies. Even though EU rules impose certain limitations on procurement procedures and price considerations often outweigh sustainability efforts, we believe sustainability efforts are winning an increasingly higher priority. We observe that sustainability has been playing a greater role among all policy considerations. Moreover, thanks to the constant inclusion and reference to sustainability, international standards are becoming both part of Copenhagen's contractual terms and also affecting the way private corporations conduct themselves.

Notes

1 The article is a contribution to CEVIA's project on Public-Private Enterprise Liability funded by the Danish Research Council for Independent Research.
2 https://ec.europa.eu/growth/single-market/public-procurement_en.
3 Environment Ministry data, available at http://mst.dk/erhverv/groen-virksomhed/baeredygtighed/groenne-indkoeb/om-baeredygtige-indkoeb/.http://mst.dk/erhverv/groen-virksomhed/baeredygtighed/groenne-indkoeb/om-baeredygtige-indkoeb/.
4 For more on the sustainability goals, please see Copenhagen: Solutions for Sustainable Cities, accessed online: https://international.kk.dk/sites/international.kk.dk/files/Copenhagen%20Solutions%20for%20Sustainable%20cities.pdf
5 See inter alia: initiatives on organic and conventional public food procurement for youth in Denmark: BioFors Report Vol.3 No 40 2008 available at www.bioforsk.no/ikbViewer/Content/34890/PDF%20NATIONAL%20REPORT%20DENMARK.pdf; Danish Timber Procurement Policy, summary available in English at www.sustainable-timber-action.org/fileadmin/files/STA_Toolkit/Buying_Sustainable_Timber_-_STA_Guide__Nov_2013.pdf; in Danish available at https://dk.fsc.org/preview.den-offentlige-indkbsvejledning.a-1595.pdfhttps://dk.fsc.org/preview.den-offentlige-indkbsvejledning.a-1595.pdf. S. Troels Poulsen, "Miljø og sociale hensyn

ved offentlige anskaffelser" in S. Treumer (eds) *Udbudsloven* (Ex Tuto Publishing, 2016); C. Risvig Hamer, *Grundlæggende udbudsret*, (Jurist og Økonomforbundets Forlag 2016) chapter 18.

6 In addition, in recent years SKI started to establish binding framework agreements for contracting authorities that want to be part of their framework agreements.

7 The Danish Competition and Consumer Authority does not have other competences or sanctions beyond bringing proceedings to the Complaints Board, which it only has done five times. It also is responsible for legislation in the field of public procurement in Denmark.

8 State Procurement & Modernisation Agency, available at www.statensindkob.dk/ServiceMenu/In-English.

9 City of Copenhagen, 2011: Procurement policy 2011–2014 – Appendix 1: Fælles indkøb – mere velfærd – Indkøbspolitik for Københavns Kommune 2011–2014.

10 See section 3.2.

11 Available at www.gronneindkob.dk/teknik-og-miljoeborgmester-morten-kabell-byerne-traekker-i-arbejdstoejet-rundt-omkring-i-verden/.

12 Communication from the Commission to the European Parliament, the Council, the European Economic and Social Committee and the Committee of the Regions: *Public procurement for a better environment*. Brussels 16.7.2008 COM (2008) 400 final.

13 Strategy on Intelligent Public Procurement (2013) available at www.statensindkob.dk/Statens-Indkobsprogram/~/media/Files/Andre%20filer/Strategi%20for%20intelligent%20offentligt%20indk%C3%B8b.ashx.

14 Available online at www.gronneindkob.dk/.

15 Nordic Council of Ministers, *Circular Public Procurement* (2017) at 15, available at www.cri.dk/sites/cri.dk/files/dokumenter/artikler/circular_public_procurement_in_the_nordic_countries.pdf.

16 Københavns Kommune, (2012). *CPH 2025 Climate Plan*.

17 For more on LCC, see Hauschild (2017).

18 C. Risvig Hamer, *Grundlæggende udbudsret*, (Jurist og Økonomforbundets Forlag 2016) at 672.

19 Denmark has ratified the ILO convention nr. 94 regarding labor clauses by *Circular on Labor Clauses in Public Contracts*, with the purpose of ensuring common wage and labor conditions for employees of enterprises that provide services to public authorities and contracting authorities. On the basis of the Circular: "All central authorities (ministries, executive agencies, etc.) must apply labour clauses . . . in all contracts for construction and civil engineering works, manufacture and production or provision of services irrespective of the amount of the contract sum." Available in Danish at uk.bm.dk/~/media/BEM/Files/ . . . /Cirkular%20on%20labour%20clauses%20pdf.ashx.

20 *Circular on Labor Clauses in Public Contracts* (in Danish: Jf. Beskæftigelsesministeriets cirkulære af 30. juni 2014 om arbejdsklausuler.).

21 Besides Copenhagen, many contracting authorities include environmental and social aspects in the competition for the contract.

22 The 10 Principles are derived from The Universal Declaration of Human Rights, including the ILO Declaration on Fundamental Principles and Rights at Work, the Rio Declaration on Environment and Development, and the UN Convention against Corruption. UN Global Compact is available at www.unglobalcompact.org.

23 UN Guiding Principles for Business and Human Rights are available at www.ohchr.org/Documents/Publications/GuidingPrinciplesBusinessHR_EN.pdf

24 The municipality of Copenhagen' CSR annex is available in English at www.kk.dk/artikel/hvilke-sociale-og-etiske-krav-stilles-i-kommunens-kontrakter.

25 The SKI's CSR annex is available in English at www.ethics.dk/asp3/tender/ski_1608_20140625.nsf/stddocs/43875A879DF62070C1257D470028D270/$file/16%2008%20New%20Annex%20E%20CSR.pdf.

26 Available at www.statensindkob.dk/ServiceMenu/In-English.

27 Available at http://oko.ku.dk/indkoebspolitik/english_version/University_of_Copenhagen_Purchasing_Policy_E.doc.
28 Available at http://greencampus.ku.dk/strategy2020/.
29 Available at www.iaruni.org/.
30 The newest revised package of the EU Public Procurement Directives came into effect in 2016: the Directive 2014/24/EU of the European Parliament and of the Council of February 26, 2014 on public procurement and repealing Directive 2004/18/EC; Directive 2014/23/EU of the European Parliament and of the Council of February 26, 2014 on the award of concession contracts.
31 "A work" means the outcome of building or civil engineering project taken as a whole which is sufficient in itself to fulfill an economic or technical function; see Article 3, of the Directive 2014/24/EU.
32 For specifics, see Annex III of the Directive 2014/24/EU.
33 For specifics, see Annex XIV of the Directive 2014/24/EU.
34 https://ec.europa.eu/growth/single-market/public-procurement_en.
35 The EU Commission admits the fact that during the consultation process, many businesses showed a general reluctance to the idea of using public procurement in support of other policy objectives – Proposal for a Directive of the European Parliament and of the Council on Public Procurement COM (2011) 896 final, p. 5.
36 Case C-513/99 Concordia Bus Finland [2002] ECR I-3609.
37 Case C-513/99 Concordia Bus Finland [2002] ECR I-3609.
38 Case C-448/01 *EVN and Wienstrom* [2003] ECR I-14527.
39 Public Contracts Regulations (2015) Commentary, Regulation 70 – Conditions for performance of contracts, available at http://pcr2015.uk/regulations/regulation-70-conditions-for-performance-of-contracts/.
40 Case C-368/10 *Commission v. Kingdom of the Netherlands* [2012] ECR I-284.
41 *Ibid*, paras 89–92.
42 Available at https://danwatch.dk/en/undersoegelse/massive-pollution-at-carlsberg-brewery-in-nepal/.
43 There are several methodological failures and reliance on biased political advocacy NGOs, including Who Profits, ICAHD and Al Haq.
44 In 2013, the Danish Ministry of Foreign Affairs stated that DanWatch received a grant of €671,000 under the CSR fund for a project to "Strengthen a watchdog who shares knowledge and creates balanced debate." DanWatch also received €9,000 grant from the Danish Ministry of Culture.
45 According to DanWatch itself, it is financed by financial contributions from private persons or funds. However, the financing of DanWatch continues to be questioned as a very little transparency is disclosed.
46 A. Piening "Best Practices & Challenges in Denmark" at the conference *Procurement beyond price: Sustainability and CSR in public purchasing*, May 4–5, 2017 in Copenhagen.
47 Outcomes of discussions at the conference *Procurement beyond price: sustainability and CSR in public purchasing*, May 4–5, 2017 in Copenhagen.
48 Often the argument against terminating a contract on the basis of sustainable clauses violations is used, stating that for the suppliers to change their behavior, guidance and collaboration from the contracting authority is needed rather than the termination of a contract.

References

Andrecka, M. (2017). "Corporate Social Responsibility and Sustainability in Danish Public Procurement" *European Procurement and Public Private Law Review* 3, 333–345.

Andrecka, M. & K. Peterkova Mitkidis. (2017). "Sustainability Requirements in EU Public and Private Procurement – A Right or an Obligation?" *Nordic Journal of Commercial Law* 1, 56–87.

Chrintz, T. (2011). *Green Procurement in the Public Sector: Potentials and Barriers*. Frederiksberg: CONCITO.

Copenhagen Council. (2009). *CPH 2025 Climate Plan: A Green, Smart and Carbon Neutral City*. Copenhagen: Copenhagen Council.

Hauschild, M.Z. et al. (2017). *Life Cycle Assessment: Theory and Practice*. Berlin: Springer.

McCrudden, C. (2007). "Corporate Social Responsibility and Public Procurement" in D. Mcbarnet, A. Voiculescu & T. Campbell (eds.), *The New Corporate Accountability: Corporate Social Responsibility and the Law*. Cambridge: Cambridge University Press.

OECD. (2009). *Territorial Review Copenhagen*. Paris: OECD.

OECD. (2016). *OECD Economic Surveys, Denmark*. Paris: OECD.

Semple, A. (2016). "The Link to the Subject Matter: A Glass Ceiling for Sustainable Public Contracts?" in Beate Sjåfjell and Anja Wiesbrock (eds.), *Sustainable Public Procurement Under EU Law – New Perspectives on the State as Stakeholder*. Cambridge: Cambridge University Press.

WHO. (2013). *Global Burden of Disease Study*. World Health Organization. Retrieved from: www.healthdata.org/gbd.

7 Innovations in public participation for sustainability and resilience in Hanoi

Linda J. Yarr and Nguyễn Ngọc Lý

Hanoi as hub

In his introductory Chapter 1 of this volume, Robert W. Orttung asks us to consider whether and how capital cities live up to their potential as hubs that aggregate, consolidate, and disseminate innovation to advance sustainability.[1] Hanoi offers a distinctive test. On the one hand, as the capital of a unitary state in which policy is closely controlled by the party-state apparatus, Hanoi enjoys a near monopoly of political and symbolic power (Logan, 2005/2006). On the other hand, its economic status is eclipsed by its southern rival, Ho Chi Minh City. And yet Hanoi has benefited from international investment and government spending that might not have accrued to it if it were not for its place as the nation's capital. Since 1975 and reunification of the country after decades of war, Hanoi has parlayed its role as political leader and seat of military control into vastly improving its economic prospects and growing role in regional and international relations. How has the city leveraged its role as a hub to effect this turnaround? With its booming dynamic economy, can Hanoi further improve the health and living conditions of its citizens and address the challenges of climate change, increased pollution, energy inefficiency, wasteful water usage, and other practices that preclude a sustainable, equitable, and resilient future?

Hanoi is an ancient city and important site of historical and cultural heritage for the Vietnamese, and indeed the world. Emerging from subjection to French colonial rule and decades of war to establish national unity and a socialist form of government, Hanoi is the seat of the Communist Party and the national institutions of the Socialist Republic of Vietnam. Even though Vietnam is now in the ranks of middle-income countries, memory of the hard-fought struggle for independence continues to motivate the leadership. Whereas the Party and government will brook no challenge to their leadership,[2] they have exercised pragmatism by shifting to a largely market economy, a reform dubbed *đội mới*, and diversifying their foreign relations, including a comprehensive partnership with the United States. This pragmatism is reflected, at least partially, in government responses to citizen activism on environmental issues.

Although other cities carry more weight in the economy and international trade,[3] Hanoi is the center for national public administration, law-making through the National Assembly, site of the country representative offices of international organizations such as the World Bank, the Asian Development Bank, the International Labour Organization, agencies of the United Nations, and the offices of influential international private voluntary organizations such as the International Union for the Conservation of Nature (IUCN), Cooperative for Assistance and Relief Everywhere (CARE), and the World Wildlife Fund (WWF). Consequently, policies and practices initiated in Hanoi influence national rules and regulations, and are in turn affected by inputs from international organizations and global norms.

While Vietnam is rapidly pursuing economic growth and advancing up the global value chain towards a more diversified economy,[4] the country faces critical environmental challenges. First, the 2018 Climate Risk Index issued by Germanwatch lists Vietnam in fifth place among the ten most vulnerable countries, citing its extreme droughts and frequent cyclones and storms resulting in flooding, salinization of the Mekong River, fatalities, and millions of dollars in property damage (Eckstein et al., 2017). Other problems include more extensive ranges for vector-borne diseases such as dengue and malaria (Chaudhry and Ruysschaert, 2017). Second, the push toward rapid economic growth has resulted in inadequate guards against pollution and environmental damages. Vietnam's air is highly polluted. Predictions from January 2017 show that "by 2030, Vietnam will be the ASEAN country most affected by coal pollution" thanks to a "premature mortality rate due to coal plant emissions, with 188.8 excess deaths per million people" (Nguyen, 2017). Currently, traffic accidents are the leading cause of unnatural deaths in Vietnam with an average fatality rate of around 106 excess deaths per million people over the past five years, according to the report. Hanoi's $PM_{2.5}$ pollution rate is "five times higher than the World Health Organization's (WHO) annual average level."[5] Toxic spills are the most dramatic evidence of poor environmental safeguards,[6] but they also point to a diminished "culture of safety," and the concern that health and safety are compromised on an ongoing basis. Attempts to respond to these critical issues are illustrative of Hanoi's position as a hub for addressing problems of sustainability and resilience, as well as its influence on programs across the country.

This chapter examines how the hub works from the perspective of an insider participant-observer, Nguyễn Ngọc Lý, the founder and director of the Center for Environment and Community Research (CECR), who is actively working on protecting the environment in the city of Hanoi, and an American academic, Linda J. Yarr. The team combines the benefits of on-the-ground knowledge of events and connections among hub components with an outsider perspective. The next section of this chapter lays out an overview of the capital city hub in Hanoi. Two case studies then examine efforts to protect the city's lakes and its tree canopy. Subsequent discussions examine efforts to spread Hanoi's best practices to other cities and attempts to measure the overall results.

Spokes of sustainability

Hanoi's capital city hub is well developed. This section surveys the key components and their interactions particularly in the areas of lakes, ponds, and canopy management.

Government agencies

In line with the United Nations' 1992 Rio Declaration on Environment and Development (Agenda 21), Vietnam's Prime Minister approved policy No. 153/2004//QD-TTg on August 17, 2004, which the National Assembly promulgated a year later as the Law on Environment Protection. Since then, the law has been amended twice, in 2005 and 2015, to address the country's environmental challenges due to population growth, rapid industrialization, and urbanization.

At the central level, in the executive branch, the Ministry of Environment and Natural Resources (MONRE) is responsible for managing land, water resources, mineral resources, and other related environmental matters. The ministry is responsible for drawing up laws, policies, and regulations to address environmental concerns.

The executive branch at the provincial level, the Provincial People's Committee (PPC), normally has 11 members appointed by the People's Council, who are in turn elected by citizens of the province. Hanoi is a special city equivalent to a province and its full name is Hanoi City, the Capital of Vietnam. The only other special city is Ho Chi Minh City. These cities are also called central cities. Other central cities are Can Tho, Hai Phong, and Da Nang, which are central cities, but not special cities like Hanoi and Ho Chi Minh City.

The Hanoi Department of Environment and Natural Resources is responsible for implementing management functions similar to those of MONRE, but at the provincial level. On substantive matters, the Department is supervised and monitored by MONRE. As the capital city, Hanoi is unique in the country since it hosts both the ministerial and provincial levels of the decision-making process.

Other ministries responsible for environmental protection include the Ministry of Construction for the drainage system and the Ministry of Agriculture and Rural Development for flood and storm control, forestry, and wildlife. These ministries have equivalent provincial departments that work at the provincial level, and thus operate in Hanoi. Other important decision-making institutions for the environment – such as the National Assembly's Committee on Science, Technology, and Environment – are also located in Hanoi.

Vietnam is a socialist country with a system of single-party rule under the Communist Party of Vietnam. While the party does not have executive power, it issued Resolution 41/2004 on the strategic direction for environment protection in the era of industrialization and modernization, which currently undergirds the decision-making process of the National Assembly and government.

Hanoi hosts the environmental decision-making bodies at all political and executive levels. On the one hand, since any environmental issue that happened in Hanoi garners quick attention at the highest levels, one would expect a robust top-down approach. On the other hand, as the environment is a complex system affecting the responsibilities of myriad stakeholders, sometimes the remits of management agencies overlap. For example, as Hanoi has been built on a wetland area of the Red River, it has a unique system of hundreds of lakes and ponds located within urban residential areas. This system of lakes and ponds offers great environmental value, helping the city to regulate water flow during periods of flooding and regulate the heat during extreme heat events. Moreover, the lakes and ponds in Hanoi also have historical significance and serve as the inspiration for artistic, religious, and cultural values.

But there is a challenge in managing this system. Historically, not all lakes and ponds were encompassed within the urban area. When Hanoi was smaller, many lakes and ponds were devoted to fish farming and therefore came under the management of the Department of Agriculture and Rural Development. As urbanization took off, within a decade or two, most of these lakes were surrounded by city dwellers, and yet they retained their fish-farming function. Meanwhile, the Department of Construction is responsible for managing lakes and ponds for drainage purposes. The Department of Environment, however, is in charge of the condition of the landscape and lakefronts. The resulting confusion and neglect of responsibilities by competing agencies has led to encroachment on – and pollution of – the lakes and ponds in Hanoi, detracting from their beauty and diminishing their environmental services.

The research community

Hanoi hosts Vietnam's best universities specialized in environmental and sustainability studies, such as Hanoi Science and Technology University, the University of Construction, and the University of Irrigation. Moreover, there are many research institutes under various ministries that carry out research in environmental matters and sustainability, such as the Central Institute for Natural Resources and Environmental Studies, the Economic and Ecosystem Institute, and units within the Vietnamese Academy of Social Sciences. These employ cadres of specialists, who engage actively in policy analysis and debate, issue research reports at the behest of the government, and interact widely with their international counterparts.

Embassies

Eighty embassies accredited to Vietnam are located in Hanoi. These embassies serve as the entry points of knowledge and information transfer between the international community and Vietnam, thereby helping to raise public awareness on certain issues of sustainability. The embassies of Denmark and Germany, for example, promoted water pollution control and sustainable wastewater treatment through a development cooperation program with the Vietnamese government and other stakeholders.

Due to their close proximity, advocacy NGOs and the research community can exchange interests and ideas directly with embassy personnel and can seek their financial and technical support. Through their grant and technical assistance programs, embassies and their affiliated donor agencies support many sustainability projects initiated by NGOs, research institutes, or the public as in the case of Hanoi Lake Report funded by the Embassy of the Czech Republic.

International organizations and international private voluntary organizations

In their 2012 report for The Asia Foundation, *A Comparative Study of Civil Society Organizations in Hanoi and Ho Chi Minh City*, William Taylor, Nguyễn Thu Hằng, Phạm Quang Tú, and Huỳnh Thị Ngọc Tuyết, found that international donors play an important role in supporting Vietnamese civil society, both technically and financially.[7] Fifty-three organizations in the sample received some form of international funding. Of those, 34 said they would feel a large impact or have an organizational crisis if international funding were cut. NGOs that have received international funds are appreciative of the positive impact of international donors. Hanoi is the site of representative offices of major international donors such as The Asia Foundation, as well as international organizations such as the United Nations Development Program (UNDP), the International Labor Organization (ILO), World Bank, and Asian Development Bank (ADB).

From 1998–2000, the UNDP in Hanoi organized regular monthly informal get-togethers with NGOs working in the environmental sector. These attracted not only international organizations, but also many activists and local NGOs based in Hanoi. While there is no formal study of the impacts of these meetings, the model was a good learning example to follow. Local NGOs established the practice of forming working groups and networks in different interest areas to share ideas and information. IUCN, one of the leading international organizations, has a project, "Promoting Active Participation of Civil Society in Environmental Governance," to support a few NGOs in Hanoi. The Center for Environment and Community Research (CECR), where one of the authors of this chapter works, has directly benefited from this project by receiving a small grant to support a community-based training program to protect lakes in Hanoi.

Civil society organizations

The aforementioned Asia Foundation report of 2012 identified 607 civil society organizations (CSOs) concentrated in Hanoi.[8] CSOs in Hanoi tend to concentrate on research, policy analysis, and technical assistance, mostly for advocacy purposes. They are relatively focused on issues related to the environment, governance, and natural resource management.

However, the capacity of CSOs is rather weak and limited. First, according to the Asia Foundation survey,[9] Vietnamese organizations are small, and their staffs mostly young and inexperienced. The average number of employees per organization is 12, and the majority of organizations polled in the survey are less than six years old. Secondly, the suspension of the Law of Associations has led to a lack of a financial mechanism for local NGOs to sustain themselves. A majority of NGOs in Vietnam depend heavily on international donors, thus their activities are dependent upon donors' priorities.

Media

A large number of national and provincial media organizations are concentrated in Hanoi, and are quick to respond to and disseminate news to the public. In both case studies under consideration here, the media played important, though different, roles in facilitating public engagement on sustainability issues. In one case, they helped draw public attention to the issue and disseminate the results of the lake research. In the other case, the media engaged actively in both raising the issue and facilitating direct actions to forestall the tree removal. It should also be noted that besides the traditional form of media such as national press or TV channels, social media is increasingly playing a role in sharing information and connecting the public, as in the case of the tree removal and planting movement.

Hub of innovation for sustainability

Responding to civil society demands for environmental protection and improved resilience to climate change, the Vietnamese government has initiated new opportunities for public participation and engagement in the sustainability policy-making process. This section describes two specific cases: 1) conservation of lakes and ponds for flood control, public health, and community enjoyment; and 2) efforts to mitigate urban heat island effects due to global warming by tree planting.

Lake and pond conservation

The shift to recognizing the utility of public participation came about over a period of seven years, from 2010–2017. It was the result of a multi-pronged approach by local citizens and newly emerged civil society organizations.

On the question of lakes and ponds, CSOs launched numerous local initiatives to advocate for community participation and to take an ecosystem approach to conserve bodies of water in Hanoi as a climate change adaptation measure. In 2010, CECR published a baseline data report on the status of 120 lakes and ponds in the six core districts of Hanoi (CECR, 2010). The report indicated that 71 percent of the lakes were heavily polluted and attributed the pollution to rapid and uncontrolled urbanization. In 2015, CECR published a follow-up report, updating the original survey of the status of Hanoi's lakes and ponds (CECR, 2015). This report indicated that since 2010, 17 lakes had disappeared or decreased in size due to pollution and encroachment by development.

The initial report attracted a wave of media attention to the situation. The average number of articles per year on lakes in Hanoi was approximately 9–10 during 2005–2009, increasing substantially to 35 articles per year from 2010 to 2017. During the seven years between 2010 and 2017, thanks to the two reports and the many activities CECR initiated about the lakes, 13 television stations featured news items, short documentaries and interviews. Outlets included both national and local TV stations: national TV channels 1, 2, 3, 4, 5, and 6;

Vietnam News Agency TV, National Assembly TV, VTC, and Hanoi TV. The VOV, the main national radio channel, also broadcast news about the reports, statistics, and activities. Meanwhile, some 88 newspapers (both print and online) published stories and articles about the lakes. An example of the type of coverage CECR received is exemplified in this item disseminated on April 13, 2017, both online and via national TV channel VTV1:

> The CECR study showed that during the five years from 2010 to 2015, 17 lakes and ponds in Hanoi have been fully reclaimed [meaning covered by land]. Total lake water surface area was reduced by 72,000 m^2. This means that a lake the size of 1.5 Thanh Cong Lakes has disappeared. This was of grave concern as this number has been recorded based on a survey of 6 core districts in Hanoi. It is very important for the city to address this issue in a comprehensive way.[10]

CECR attributes this uptick in media attention at least partly to the Hanoi lakes reports. The reports also increased the spread of the ecosystem approach to urban lake management. Before 2010, articles about Hanoi's lakes and ponds were fragmented reports emphasizing the incidence of pollution. After 2010, analyses reflected a more comprehensive understanding of lakes management and highlighted inadequacies and shortcomings in current institutional practices.

Both reports were valuable reference documents for use by Hanoi's Construction Department, the Hanoi Environmental Protection Agency, and Hanoi's Institute of Planning and Construction. As a result, a series of national and international conferences were organized to discuss the role of urban lakes in the context of the rapid urbanization of Hanoi and challenges of urban lake management. In 2012, CECR organized an international conference on the "Sustainable use of Hanoi's lakes with public participation in the context of urbanization and climate change" featuring 18 presentations on sustainable lake management and public participation (CECR, 2012). Among the 110 attendees were representatives of a variety of institutions, including universities, foreign embassies, and government agencies. The Hungarian embassy invited a prominent expert from Hungary to address the audience, and the Ambassador of the United States opened the conference alongside the Minister of Science and Technology. The conference itself was a manifestation of the hub function of a capital city.

That year also saw the establishment of the Hanoi Lake Club, a community group dedicated to protecting the lakes and raising popular awareness of the problem. CECR and the Institute for Urban Studies and Infrastructure Development sponsored the Hanoi Lake Club. From 2012–2017, the Hanoi Lake Club organized 13 major lake cleanup events. These events attracted more than 10,000 people from lakeside communities, universities, local volunteer groups, and businesses such as Lock & Lock, Standard Chartered, and others.

Since the founding of the Hanoi Lake Club, women and youth groups have played a central role in initiating and implementing lakes conservation campaigns. In Hoang Van Thu ward, the Women's Union launched a movement

"Green – Clean – Beautiful – Blossoming." The group installed additional trash bins around Den Lu Lake, and assigned households to manage the lakefronts near their homes. In 2012, the Women's Union started an awareness-raising campaign during the Lunar New Year Holiday to discourage people from dumping religious waste into the lake – a ritual commonly practiced in Vietnam during holidays.

The Women's Union not only initiated the campaigns, but with its unique position as a socio-political organization and its extensive communications reach, the Union also connected and mobilized other organizations, companies, and community groups to participate in lake conservation. In Quang An ward, the Women's Union worked closely with the Veterans Association and Quang An ward cooperatives, with technical support from CECR, to conduct an assessment of the conditions of Cheo and Pho Linh ponds. Based on the assessment, the Women's Union and key stakeholders devised lake conservation plans for the two lakes. The plans involved regular cleanups, dredging, shore improvement, and awareness-raising activities. Youth clubs also contributed greatly to the activities of the Hanoi Lakes Club through the monitoring and sharing of information on the water quality of the lakes, and convincing households to sign lake protection commitments.

Tree preservation

In the second case featured in this chapter, Hanoi was the site of a major movement to preserve the local tree canopy. In March 2015, an unprecedented "green movement" in Hanoi was organized to oppose the city government's scheme to cut down 6,700 trees. This movement not only attracted media attention in Vietnam, it also made headlines in the international press. The *Financial Times* declared that this movement "symbolizes wider social frustration in this authoritarian state."[11]

Under citizen pressure, the Chairman of the Hanoi People's Committee (the city government) issued a decision to stop the project, disciplined some officials, and set up an interdisciplinary inspection team.[12] This was a successful social movement, taking place peacefully, with the participation of almost every sector of the city's population.

The decision to remove 6,700 of Hanoi's trees was issued in 2013 with the aim of replacing decaying trees with new ones and making space for the construction of an urban light rail system. However, the public noticed healthy and century-old trees in areas outside of the project being cut down without justification. Mounting unease did not coalesce into a social movement until Trần Đăng Tuấn, a reputable journalist and former deputy general director of the Vietnam National Television station, filed an open letter to the Chairman of Hanoi's People's Committee voicing concerns over the project and calling for public consultation on the tree removal. The letter was published as a Facebook post and attracted 6,567 likes, 789 shares, and 588 comments. However, on March 17, 2015, when the Deputy Head of the Propaganda Committee declared in response to a journalist's question during the press briefing of the Hanoi People's Committee that the city authorities "do not have to ask the people about cutting trees," and this comment was published widely on *Vietnamnet* and most of the public media, an intensive social movement took off, both online and offline.

A series of actions to protest the felling of trees, such as "Love Trees – Save Cities," "Tree Hugs," and "Green Walk," proceeded in an organized and non-violent manner. This movement's membership was extremely diverse. Ordinary citizens, CSO staff, the media, artists, and researchers all played roles. One of the most active and pioneering groups in the movement was "Group 6700 people" which was established on Facebook by a housewife with no prior experience in social activism.[13] The initial aim of the page was to get 6,700 likes to pay tribute to the trees. However, when the page quickly received 10,000 likes within 24 hours, the group started to organize public campaigns. These actions were organized and reported online through a Facebook group using social media hashtags, which garnered further media attention. Celebrities also joined the movement through sharing comments and mobilizing their fan bases to participate. Non-governmental organizations, research institutions, and experts contributed by holding conferences in order to share information on the tree replacement project and Hanoi's urban planning. Never before had information on urban management, urban green management, and international experience in managing trees and replacing trees been more widely disseminated and received. Journalists also contributed by conducting investigative research on the cost of the project, the use of timber from the cut-down trees, and all the actors involved in the project.[14]

Outcomes

These sets of actions resulted in the following concrete measures.

Protection of lakes and ponds

The public's new-found concern for the status of lakes and ponds in Hanoi instigated a new approach to their management and a shift in the understanding of the value of lakes and ponds. In the past, bodies of water were considered part of the drainage system rather than a public amenity and constituent of the city's ecosystem. A top-down approach in master planning prevailed and public participation was not part of the decision-making process, nor was the public included in the process of monitoring and implementation.

That mindset has changed. In 2016, the Ministry of Construction decided to prepare a new set of policy guidance entitled the Orientation for Development of Water Drainage in Urban Centers and Industrial Parks. In preparing the document, the Ministry organized a special discussion and conducted analyses of lakes and ponds in Hanoi. On the basis of recommendations proposed by CECR and specialists working on lakes and ponds, the document has been adjusted to integrate the need to conserve, develop, and effectively utilize lakes and ponds. It includes specific targets on lakes and ponds to be incorporated in subsequent documents. This significant policy officially endorses public participation in planning and monitoring.

Second, authorities organized public participation and consultation on the restoration of Hanoi's iconic Hoan Kiem Lake. In September 2017, Hoan Kiem district held a public consultation meeting at the Hoan Kiem's People Committee. Around 50 participants, community residents, members of women's union chapters, university researchers, scientific institute researchers, and ecosystem experts were invited. Leaders of civil society organizations spoke alongside academics and private sector representatives. On the basis of the comments and suggestions made during the consultation, the Hoan Kiem Lake restoration project has been revised to integrate measures to conserve biodiversity and include a process of public consultation.

Tree preservation

Expressions of public anger resulted in government investigation into the issues, and some senior staff members were disciplined. A new design for tree planting has been developed so that today Hanoi is greener and some sections of the city are now more attractive. Hanoi authorities estimated that by the end of 2015, trees density reached 6.7–6.8 m^2/person. In 2016, 600 new trees were planted daily.[15] By 2020, tree density will reach 10–11 m^2/person, according to the tree management plan published by the Hanoi government.[16] The planting, watering, and management of trees are the sole responsibility of Hanoi Green Park Company, which operates under the supervision of Hanoi's Department of Construction.

The city authorities also publicly responded to concerns of the people and media over storm damage to trees. For the first time, in July 2017, the Chairman of Hanoi's People's Committee, a local state administrative agency that is elected by the local Party authority, namely The People's Council, published the total number of trees being cut down or moved during the period 2015–2017. According to the People's Committee's report, during this period 813 trees were cut down and 832 were replaced. The city authorities also published the tree removal plan for the period 2017–2018.

Since 2015, the city authorities have also conducted public consultations more frequently when it comes to construction projects that will potentially result in tree removals. For example, in May 2017, Hanoi's Construction Department conducted a public consultation on the removal of 1,300 trees for the extension of Ring Road 3. That month, Hanoi's Construction Department also opened to public comment a plan to remove 4,000 trees in central Hanoi. The city has still not proceeded with the proposed removal. Consequently, during 2016–2017 there was a shift in the attitude of Hanoi's authorities towards public participation in urban tree management, thanks largely to the social movement to deter tree cutting.

Obstacles

In the two cases under study, one can identify several obstacles to public participation. First, there is a very limited legal framework to encourage and facilitate public participation. Article 146 of the Environmental Protection Law specifies the rights and duties of the community. Accordingly, the community has to be consulted during environmental impact assessments and has the right to demand environment-related information from businesses. However, this has not been translated into follow-up decrees and circulars, nor has funding been made available to implement this article. As a result, public participation is limited to commenting on environment protection projects. Meaningful public participation in the decision-making process is limited.

Another obstacle to public participation on conservation of the lakes is insufficient funding to sustain community activities on behalf of the lakes. Communities rely on their own funding, thus limiting the scale of the protection projects. Although the cleanup activities are encouraged and appreciated by the city authorities, the support has not been translated into earmarked funding for community participation and public outreach.

Furthermore, the capacity of civil society organizations is rather weak and limited. Vietnamese NGOs are small with inexperienced staff. There are few domestic funding sources for the groups that do exist, forcing them to rely on international donors and their priority preferences.

Convergence in the hub and move to innovate

In the case of Hanoi's lakes and ponds, it was CECR, a local NGO, that spurred public interest in the protection of Hanoi lakes. This chapter's co-author, Nguyễn Ngọc Lý, after reading about how Hanoi lakes were dying, sought to find out just how many lakes there were. Discovering that no data was available, she and her team solicited help from other NGOs, research institutes and government officials to collect baseline data on the number and status of Hanoi's many lakes.

The two baseline reports they produced highlighted the severe trend toward increased pollution of all the lakes in Hanoi and urgently called for an ecosystem approach with public participation in the protection of urban lakes. The reports raised the awareness of communities, researchers, and policymakers on the issues, and recommended concrete actions for each of the stakeholders to take in order to protect Hanoi's lakes. Mass organizations such as the Women's Union also had an important role to play. The Women's Union at the ward level initiated cleanup activities and awareness-raising campaigns. With the capacity-building support of local NGOs, the Women's Union also conducted pollution assessments and worked closely with the Veterans' Association and People's Committee to devise a lake monitoring and protection plan.

As mentioned earlier, Trần Đăng Tuấn, a well-known journalist with a large number of followers on social media, sparked the tree protests. However, many actors contributed to the social movement in their own way. While university students and celebrities mobilized people to engage in direct campaigning, such as "Tree Hugs" and "Yellow Ribbons: Love Trees – Save City," NGOs like Pan Nature and the Center for Media in Educating Community (MEC), experts, and intellectuals held conferences and wrote analyses. For example, architect Trần Huy Ánh, in an interview with *The Laborer* newspaper, suggested that "New and small trees can be planted next to old trees so as to keep the air conditioning function of the tree canopy. Cutting down green trees clearly affects the environment of the capital."[17] Well-known mathematician Ngô Bảo Châu, formerly of the University of Chicago, also posted on social media a series of questions pointing towards the controversies and lack of transparency of the tree removal project[18] to share information, identify inadequacies in the tree removal plan, and propose best practices in urban tree management. The media also played an important role in the movement through their investigative research and constant reporting on the opinions and actions of city residents.[19]

IUCN, one of the leading international organizations, has a project "Promoting Active Participation of Civil Society in Environmental Governance" to support Hanoi's handful of environmental NGOs. CECR has directly benefited from this project by receiving a small fund to support community-based training programs to protect Hanoi Lakes.

The Lake project received strong support from various foreign embassies in Hanoi: The Czech Embassy provided financial support to collect information from lakes in 2010. The U.S. Embassy provided an advisor to work with CECR to develop a model of community-based lake management and presented the case of Lake Tahoe at the International Conference on Hanoi Lake organized by CECR in 2012. The Hungarian Embassy also provided support by sending an expert from Hungary to the same conference to present the case of Balaton Lake as an example for consideration for Hanoi. The U.S. Embassy joined CECR in organizing Hanoi's first Earth Day celebration on April 22, 2012, attracting strong participation by Hanoi citizens. Since then, Earth Day has been organized annually by the U.S. Embassy, CECR, and the Vietnam Environmental Administration, with businesses, NGOs, students, and women's groups in Hanoi taking part to campaign for water protection.

Though not directly, universities and research institutes contributed to public participation in lakes protection and the tree protection movement by providing research and technical support to the public. The Hanoi Lake Reports, for example, were made possible due to collaboration among different actors including the NGO CECR, universities, and research institutes. To analyse water samples, CECR collaborated with Hanoi Science and Technology University. Hanoi University of Industrial Arts assisted with taking pictures of the lakes. Data and information about the lakes was represented by Geographic Information Systems (GIS) with technical support from the Institute of Geophysics. The Hanoi Lakes Report may have been the idea of a single organization, but once the idea caught on, it was realized by the collaboration of a network of various actors, including foreign entities. For example, the Czech Embassy provided the funding for the research and report.

Dissemination beyond the capital city

Protection of lakes and ponds

The movement to protect water resources spread to other cities. After the first Earth Day cleanup events held in Hanoi in 2012, Earth Day was also organized in Ho Chi Minh City in 2014 by the Center for Skilled Education and Community Communication, a local NGO. Another local NGO, the Coalition for Clean Water, initiated public participation in water resource protection. The Coalition is a network of NGOs, research institutions, policymakers, experts, and women in six provinces of Vietnam. In Da Nang and Binh Duong, the Coalition helped to organize public consultations to garner public inputs to water protection plans. In these areas, it also formed core groups of community residents to monitor water quality and regulation compliance. These groups receive training and support from local NGOs to carry out monitoring and restoration of water resources.

Preservation of trees

To date, citizen activism to preserve the tree canopy has not been replicated in other cities.

Measuring results

Protection of lakes and ponds

In the last three years, the Hanoi authorities have decided to improve water quality by dredging and improving the protection of lakes. So far, they have successfully dredged large lakes such as Hoan Kiem and Ngoc Khanh. A separate drainage system has been added to collect wastewater and rainwater. Over the coming year, the West Lake will be dredged and rehabilitated. Fountains have been installed in most of the lakes to improve aeration, and aquatic plantation beds for pollution control. Building encroachments on the lakes or ponds have been strictly curtailed, with no further cases of illegal encroachment reported. The government has officially organized public participation for projects related to the two major lakes, Hoan Kiem and West Lake.

Preservation of trees

Surveys of eight newspapers, most of which are online, revealed that 373 articles about the tree removal project were written during the period November 2014 to August 2015.[20] The majority of them covered either the city authority's plan (46.3%) or the tree removal (44.5%).

Apart from the "One Million Trees by 2020" initiative, since 2016 the city authority outsourced tree protection and planting to the Hanoi Green Park Company instead of dividing the work among several actors that would be harder for the public to monitor. The Company was given significant resources to improve the efficiency and effectiveness of tree management. The Company staff and Hanoi Construction Department officials were sent to China and Singapore to learn about tree management. These actions demonstrate that Hanoi's authorities pay a lot more attention and have become more cautious when it comes to tree management in the city.

Conclusion

In the context of Vietnam's single-party political system, where the government continues to hold considerable sway over public opinion and centralized authority over regulations governing public administration, expenditures, and public safety, the cases outlined here provide an indication of how the "hub function" of a capital city can be tapped to spur innovation toward sustainability. Under pressure from a constellation of "spokes" of the hub – civil society organizations, universities, foreign embassies, international organizations, and private voluntary organizations, businesses, and media – the Hanoi city authorities responded by instituting measures for public consultation, an innovation in what is otherwise a unitary, and some would say authoritarian, political sphere. In fact, a more serious impediment to citizen and civil society organization activism on the environment, climate change, and sustainability is the lack of a law on associations. Nevertheless, the public consultation efforts described here represent a precedent setting and, one hopes, demonstrable movement toward engaging the citizenry in making decisions that affect the environment and their well-being.

Notes

1 The authors gratefully acknowledge the superb research assistance provided by Pham Ngoc Anh, project officer, and Nguyen Thị Thu Ha, researcher, Center for Environment and Community Research (CECR).
2 Murray, Bennett, "Vietnam's harsh summer: State launches largest crackdown on dissidents in years," *The Guardian*, September 25, 2017, www.theguardian.com/world/2017/sep/26/vietnams-state-largest-crackdown-on-dissidents-years accessed November 6, 2017.
3 In May 2017, Haiphong led the country in industrial production growth. Hanoi came in fifth after Thai Nguyen, Da Nang, and Ho Chi Minh City. "Hai Phong takes lead in industrial production growth," *Nhan Dan Online*, May 29, 2017, available at http://en.nhandan.com.vn/business/economy/item/5236802-hai-phong-takes-lead-in-industrial-production-growth.html.
4 Over the last decade, the contribution of agriculture to the gross domestic product (GDP) declined by 0.59%, manufacturing by 2.21%, whereas services gained 2.8%. "Vietnam: Distribution of gross domestic product (GDP) across economic sectors from 2006 to 2016," *Statista: The Statistics Portal*, available at www.statista.com/statistics/444611/vietnam-gdp-distribution-across-economic-sectors.

5 "How serious is air pollution in Vietnam?" *VietnNamNet Bridge*, April 18, 2017. Available at http://english.vietnamnet.vn/fms/environment/176546/how-serious-is-air-pollution-in-vietnam.html.
6 Ives, Mike, "Outrage Over Fish Kill in Vietnam Simmers 6 Months Later," *The New York Times*, October 3, 2016. Accessed online November 6, 2017 at www.nytimes.com/2016/10/04/world/asia/formosa-vietnam-fish.html.
7 The Asia Foundation; Xã hội dân sự tại Việt Nam: Nghiên cứu so sánh về các tổ chức xã hội dân sự tại Hà Nội và thành phố Hồ Chí Minh; 2012 tải về từ. http://tuthiendoanhnghiep.com/Uploads/files/TAF-CSOs-Comparative-Study-Report-VN.pdf.
8 *Ibid.*
9 *Ibid.*
10 http://vtv.vn/trong-nuoc/17-ao-ho-tai-ha-noi-bien-mat-chi-sau-5-nam-20170413092937306.htm.
11 www.ft.com/content/54d07f2a-d462-11e4-8be8-00144feab7de.
12 http://news.xinhuanet.com/english/2015-03/20/c_134084601.htm.
13 Ngoc Anh Vu, Grassroots Environmental Activism in an Authoritarian Context: The Trees Movement in Vietnam; 2017; tải về từ http://opus.bath.ac.uk/54061/1/AnhVuNgoc.pdf
14 http://kienthuc.net.vn/soi-xet/cay-xanh-ha-noi-bi-chat-go-duoc-mang-di-dau-469147.html.
15 http://tbdn.com.vn/moi-ngay--ha-noi-trong-them-duoc-600-cay-moi_n26521.html
16 http://vov.vn/xa-hoi/chu-tich-tp-ha-noi-se-cong-khai-viec-danh-chuyen-chat-ha-cay-xanh-639055.vov.
17 http://nld.com.vn/thoi-su-trong-nuoc/cay-xanh-ha-noi-ton-thuong-20150319231412731.htm.
18 https://vnexpress.net/tin-tuc/cong-dong/gs-ngo-bao-chau-len-tieng-ve-viec-chat-6-700-cay-xanh-o-ha-noi-3159476.html.
19 Typical articles include: http://soha.vn/cu-dan-mang/nhung-cay-xanh-bi-chat-va-noi-long-cua-nguoi-dan-ha-noi-20150319103613927.htm; www.thesaigontimes.vn/127879/Du-luan-phan-ung-Ha-Noi-dung-ke-hoach-chat-6700-cay-xanh.html; https://baotainguyenmoitruong.vn/ban-doc/nha-bao-tran-dang-tuan-kien-nghi-ha-noi-dung-chat-bo-6–700-cay-xanh-787631.html.
20 http://isee.org.vn/Content/Home/Library/473/bao-cao-phong-trao-6700-bao-ve-cay-xanh-o-ha-noi.pdf.

References

CECR. (2010). *Report on Hanoi Lakes 2010*. Hanoi: Center for Environment and Community Research.

CECR. (2012). *Proceedings of International Conference on Community-Based Sustainable Use of Hanoi Lakes in the Context of Urbanization and Climate* Change. Hanoi: Center for Environment and Community Research.

CECR. (2015). *Report on Hanoi Lakes 2015*. Hanoi: Center for Environment and Community Research.

Chaudhry, P., and Ruysschaert, G. (2017). *Climate Change and Human Development in Vietnam: A Case Study*. Human Development Report Office Occasional Paper, Human Development Report 2007/2008, draft of 26 April 2007. Available at https://pdfs.semanticscholar.org/5091/58b31e47a9272b4e7743a491d4d6b83aa20b.pdf.

Eckstein, D., Künzel, V., and Schäfer, L. (2017). *Global Climate Risk Index 2018: Who Suffers Most from Extreme Weather Events? Weather-related Loss Events in 2016 and 1997 to 2016*. Bonn: Germanwatch, 7. Available at https://germanwatch.org/en/download/20432.pdf.

Logan, William S. (2005/2006). "The Cultural Role of Capital Cities: Hanoi and Hue, Vietnam," *Pacific Affairs* 78 (4), 559–575.

Nguyen, V.P. (2017). "With Growth of Coal Power Plants, Vietnam's Future Is Grim," *The Diplomat*, 25 January. Available at https://thediplomat.com/2017/01/with-growth-of-coal-power-plants-vietnams-future-is-grim.

8 Delivering clean water in an African capital

Jennifer Lessick

In this chapter, we turn the focus toward the global South. Africa is the world's most quickly urbanizing continent: the number of city dwellers will double in the next twenty years, making the future urban population over 654 million people, or 50 percent of the African population (Echart et al., 2012). Sustainability and innovation on this continent are crucial for maintaining and improving living standards. The demand for food, water, and energy is growing at an exponential rate, and the pressures will only continue to intensify. The importance of including a study of an African nation with significantly fewer resources, considerably weaker governance, drastically higher pressures, and greater environmental risks is critical to understanding the essential question of capital cities. How can capital cities – not just those with wealth and widespread opportunity – become role models for future sustainable innovation? This study finds an answer in Antananarivo, Madagascar.

With a population of over 1.6 million, Antananarivo, the capital commonly nicknamed "Tana," is the cultural, economic, and political center of Madagascar (WSUP, no date a). It is home to many universities, dozens of museums, the heart of the banking system, and the headquarters of the government. Because of its comparable abundance of professionals, scholars, and most importantly international networks, the city is able to innovate at a much faster pace than the rest of the country. By using these unique aspects of a capital city, Antananarivo has been able to lead the way in providing clean and safe drinking water to the citizens of Madagascar, despite the extreme climatic, social, and political challenges faced by many African nations. Through an analysis of the improved water systems in Antananarivo, this chapter will discuss ways in which Madagascar can spread these advances across the country where the hub-like characteristics of the capital do not exist.

Tana was founded in the 1600s by the Merina people who inhabited the central plateau of Madagascar for centuries. The Merina established a kingdom in the 15th century, building vast irrigation systems in order to cultivate rice, cassava, potatoes, onions, and other crops. When the Merina kings consolidated power on the island in the 17th century, they established Antananarivo and built a Royal Estate which can still be seen today. The biggest industries in the city include tobacco and food processing in addition to the manufacture of leather goods and clothing. The city is also home to many research institutes, an observatory, the Bibliothéque Nationale, and the University of Madagascar (Editors of Encyclopedia Britannica, 2018).

Antananarivo as a hub

Antananarivo presents complex problems in thinking about the capital city hub at the center of this book. As the largest city in Madagascar by great lengths, it is the undebated center of economic activity for the entire country, the seat of the federal government, and the anchor of many international and NGO investments that come into the country.

However, despite its numerous advantages, Antananarivo is not living up to its full potential as a hub. While this case study will illuminate how this capital city was able to harness international attention in order to fuel investments to a much-needed infrastructure project, it is difficult to assert that a fully functioning capital city hub had much to do with the progress achieved so far. Antananarivo has achieved a considerable amount of success in building integrated water systems, but this success is limited by the considerable lack of resources, absence of strong economic vitality, and numerous missing hub components. While there is some interaction between the international players and the local city government, the existing components of the hub are not developed enough to work together in a way that shows that they are more than a sum of their parts.

Challenges facing water supply in Sub-Saharan Africa

To place Tana's policies in context, we lay out the water challenges faced by Sub-Saharan Africa as a whole. As of 2015, only 68 percent of the population of Sub-Saharan Africa had access to improved sources of drinking water (Rondi, 2015). Water services in many cities were characterized by intermittent supply, frequent breakdowns, inefficient operations, poor maintenance, and depleted finances. These issues led to 1.6 million deaths per year, mostly due to poor sanitation which, only reached acceptable levels in 40 percent of urban areas and 23 percent of rural areas as of 2015 (Rondi, 2015). The countries with the best sewage services – South Africa, Namibia, and Senegal – have an 80 percent coverage rate (Jacobsen, 2013). Africa's rapid urbanization means that demand for water on the continent will quadruple over the next twenty-five years, according to the World Bank (Jacobsen, 2013). While 60 percent of current urban residents live in slum areas where water supply and sanitation are severely inadequate and this number is forecasted to grow by 4.53 percent per year, urban water management progress is struggling to keep up with the rate of urbanization (Jacobsen, 2013). Just over one billion people currently live in Africa, with 40 percent of those living in urban areas (Jacobsen, 2013). However, the urban population is expected to reach 1.23 billion people by 2050, with 60 percent of those living in cities (Jacobsen, 2013). Additionally, by 2030, the continent's urban population will surpass its rural population, with 48 current urban areas in Africa hosting over one million people – and two of those (Cairo and Lagos) already considered mega-cities (Jacobsen, 2013). Clearly, African policy makers need to focus on how they plan to provide water supply and sanitation to this rapidly growing population.

In addition to the challenges of rapid urbanization, the lack of political will in many African countries contributes to the water coverage issues seen today. Many leaders have not been able to allocate national resources to undertake the necessary reforms to improve and expand water systems, failing to meet annual estimated infrastructure spending needs for water supply of $14.9 billion and water supply operation and maintenance needs estimated at $7 billion per year (Jacobsen, 2013). This lack of resource distribution to the water sector lets fester aging infrastructure that contributes to water losses of more than 30–50 percent of supplies, a lack of regular maintenance, a poor understanding of what needs to be done, and emerging contaminants which creates widespread consumer discontent (Jacobsen, 2013). Many customers are forced to take matters into their own hands by resorting to unsafe sources of water or boiling tap water due to lost confidence in the utilities. All of these issues are intensifying due to the increasing impact from climate change, and the situation is likely to continue to deteriorate.

Arguably the most pressing issue for Africa, especially Sub-Saharan Africa, is the fact that the region is one of the most vulnerable areas in the world when it comes to the impacts of climate change. Forty-five percent of Africa is considered arid or extremely arid, while 22 percent is semi-arid (Jacobsen, 2013). These water-scarce areas will experience increased and strengthened droughts as climate change worsens. In the past twenty years alone, the available fresh water resources in Africa shrank significantly due to severe and prolonged droughts, a situation that will only get worse as the effects of climate change heighten (Collier et al., 2008). Meanwhile, 33 percent of Africa is sub-humid or humid (Jacobsen, 2013). This means that the excess precipitation exacerbated by climate change will generate high loads of water and pollutants that will contaminate and endanger the use of those remaining available resources. In light of these expectations, the Population Action International Study projects that by 2025, fifteen African countries will face water scarcity, while another eleven will be water stressed (Jacobsen, 2013).

Integrated Urban Water Management in Africa

The World Bank's Integrated Urban Water Management (IUWM) plan seeks to provide clean water and sanitation in a manner which will promote sustainability and resiliency. IUWM categorizes African cities based on their level of water supply and sanitation management in order to encourage better performance. The World Bank IUWM mission statement asserts that:

> by adopting integrated urban waste management approaches, policy makers in African cities have a real chance to address diverse issues such as increased competition for water with upstream water users, improve urban planning

> by understanding water's interaction with other sectors, and in the face of a changing climate, secure resilience in an uncertain future by relying on a diversity of water sources.
>
> (Echart et al., 2012)

The aim of IUWM is to urge African policy makers to look at water management as a holistic system that includes all components of the urban water cycle. It requires attention not only to providing water, but to water scarcity, water quality, and energy use. The plan encompasses everything from agricultural techniques to ethical business practices, from government policy-making to public education and understanding, and much more. The World Bank cites some successful IUWM approaches already in use: Windhoek, Namibia, where 26 percent of water supply comes from wastewater reuse; Arua, Uganda, where decentralized wastewater treatment systems and soil aquifer treatment is expanding to the outskirts of the city; and Nairobi, Kenya, where leakage reduction, storm and rain water harvesting, and graywater recycling are promoting resiliency in the city (Echart et al., 2012).

The stages of urban water management (UWM) development are usually broken down into six distinct categories: the water supply city, the sewered city, the drained city, the waterways city, the water cycle city, and the water sensitive city (Echart et al., 2012). Water supply cities refer to those with only supply hydraulics, sewered cities are those that also have separate sewerage schemes, drained cities include drainage and channelization, waterways cities incorporate point and diffuse source pollution management, water cycle cities have diverse water sources and promotion of waterway protection, and water sensitive cities are adaptive and include multi-functional infrastructure and urban design in order to reinforce water sensitive behaviors (Echart et al., 2012). While cities may be able to skip some stages and still exhibit characteristics of different stages, this approach is helpful in benchmarking the current state of UWM in African cities in comparative perspective.

After the colonial period in Africa, urban water management had "limited human and institutional capacity and financial constraints which have limited the capacity of utilities to expand their coverage to cope with the rapid urban growth" (Echart et al., 2012). As a result, water infrastructure greatly deteriorated. Because of the lack of proper planning and capacity, urban areas grew haphazardly with little successful infrastructure development. The result was large slum areas with little to no sewage or water supply infrastructure combined with city centers with deteriorating and disorganized water systems. As a result, most African cities made no progress in transitioning towards the level of a sewered city or drained city (Echart et al., 2012).

Today, most African cities remain stuck at the stage of water supply city. Few cities have been able to develop to the level of a waterways city and even fewer have been able to reach the stage of water cycle city, where "the UWM is driven by the limits of the natural water resources and focus in concepts of water conservation, water reclamation, cascade use of water and diversifying of water sources"

(Echart et al., 2012). Despite this, as Africa continues to rapidly urbanize, the World Bank asserts that there is a large "window of opportunity to identify efficient development trajectories that will enable it to leapfrog some of the transition steps" (Echart et al., 2012). Because of this, there is a possibility for water systems to emerge in Africa as more efficient, more sustainable, and better organized than elsewhere in the world.

Challenges facing Madagascar

How does Madagascar fare in addressing these challenges? Not well. In terms of water and sanitation coverage, Madagascar is one of poorest performers on the continent. In August 2008, twelve million people of the country's nineteen million population (63.2 percent) lacked access to potable water, and fourteen million (73.7 percent) lacked access to proper sanitation (UNDP, 2009).

This lack of coverage was largely due to three major weaknesses in Madagascar. First, poor governance, including limited access to information and human resources, is a big driver in the lack of access to water and sanitation. Second, limited institutional capacity and knowledge of effective legal frameworks creates issues for governmental organizations and international networks working with the city's water utility. Finally, limited financial resources means restricted opportunity to act; in many cases providing clean water and sanitation seems like an impossible task, particularly in slum areas (UNDP, 2009). These issues are exacerbated by the fact that Madagascar is one of the world's poorest countries, with 92 percent of the population living on less than $2 per day and 72 percent of the population living in slums (UN, 2015).

Additionally, Madagascar is the second most vulnerable country in Africa to hazardous weather events, in particular, tropical cyclones. While these storms result in significant damage and extensive loss of life, the southern regions of the country sustain a recurrent threat of drought which results in chronic and widespread malnutrition with famine recurring every few years. For example, in late 2016, a drought after three consecutive years of failed rains left the country with crop failure resulting in nearly 840,000 people experiencing a Phase 4 Integrated Food Security phase classification, with Phase 1 indicating an abundance of food and Phase 5 meaning famine (Chaterjee, 2016). Meanwhile, 330,000 people were on the brink of famine, as reported by the United Nations (Chaterjee, 2016). These climate issues will undoubtedly worsen and intensify with the increased effects of climate change forecast in the coming years. In particular, cyclone intensity could increase by 46 percent and shift northward, resulting in large effects on agriculture and food security (World Bank, 2011). In terms of water quality, the country is extremely vulnerable due to the potential of increased waterborne diseases affecting its inefficient water systems (World Bank, 2011).

It is obvious that Madagascar needs to act on these issues. In terms of the six IUWM indicators, Antananarivo, the largest city in Madagascar, is struggling. The urbanization challenge is front and center, with the degree of urbanization in the country increasing 6.31 percent in the last decade alone (Statista, 2017).

As noted earlier, this is a large issue due to the fact that 72 percent of the urbanizing population lives in slums (WSUP, no date a). Solid waste management is poor in the city, mostly due to the lack of authority over the systems currently in place. As for water resource availability, the effects from climate change are set to dramatically reduce the amount of water available to the country as a whole, which will undoubtedly affect Antananarivo in the future. Water sanitation and supply services are an issue, with a sizeable proportion of urban residents lacking access to both. Lastly, flood hazards in Antananarivo do not seem to be a pressing issue, so more attention can be paid to the other indicators discussed.

With a growing and urbanizing population, there is ample opportunity for Madagascar to implement sustainable practices which will provide water and sanitation for its citizens in ways that will remain resilient to the present and future effects of climate change. The government of Madagascar has taken the initiative over the past year to align its vision with the UN's 2015 Sustainable Development Goal 6 to "ensure availability and sustainable management of water and sanitation for all," in order to provide universal and equitable access to safe and affordable drinking water and adequate sanitation and hygiene while also improving water quality by reducing pollution, sustainably increasing water-use efficiency, and implementing integrated water resources management at all levels (UNICEF, 2016). To predict how Madagascar will be able to do this in the coming years, we can look into the past accomplishments in Antananarivo, the capital of Madagascar.

What innovation has been successful in Antananarivo?

In 2016, the City of Antananarivo, the Madagascan Ministry of Water, and the state-owned utility JIRAMA, were recognized by the FT/IFC Transformational Business Awards for Excellence in City-Led Transformation (Financial Times, 2016). These entities gained acclaim for providing 670,000 low-income residents, or one-third of the city's population, with improved water supply since 2010 by (1) creating a dedicated unit responsible for serving low-income consumers; (2) creating a dedicated unit responsible for reducing losses caused by theft and leakage, and improving payments collection; (3) increasing water capacity with relatively low-cost augmentation schemes; and (4) improving water delivery partnerships with community-based Water User Associations to ensure better service for low-income residents (WSUP, no date b). In order to understand how to bring this technology and innovation to other areas of Madagascar, it is necessary to discuss the framework and steps taken to accomplish this feat.

While Madagascar is a need-heavy country by any measure, it does not receive many funds from international donors like the United States. This neglect is largely due to a political crisis beginning in 2009 that destabilized the country, deterring international investments for some years. The crisis involved tensions between the elected president, Marc Ravalomanana, and an opposition movement led by the mayor of Antananarivo, Andry Rajoelina, which resulted in the ouster of Ravalomanana and a complete overhaul of the government, which in turn ignited protests, violence, and controversy (Ploch and Cook, 2012). When the unrest subsided, organizations like the U.S. Agency for International Development (USAID), Water and Sanitation for the Urban Poor (WSUP), and the European Union (EU) made available funding to the newly stabilized government. However, investments in Madagascar are still considerably smaller than those provided to mainland states like the Democratic Republic of the Congo, Uganda, and Rwanda. Despite this considerable disadvantage in international capital, Antananarivo was still able to make sizable accomplishments in collaboration with international non-profits (Keita, 2017).

The water improvements at the core of this case study occurred largely due to the efforts of WSUP in combination with JIRAMA. WSUP is an organization that is present in seventeen cities across Africa, Europe, and Asia, and is a unit of the United Nations that aims to advance the Sustainable Development Goals for water and sanitation. It works through multi-sector and stakeholder partnerships (most notably the government of the United Kingdom) in order to deliver sustainable, equitable, and affordable water and sanitation services to urban poor in developing countries. It targets urban areas that are marginalized by water utilities, typically slums. WSUP addresses the challenge of water and sanitation in these areas utilizing a four-pronged approach including programs, research, advice, and investment support (WSUP, no date c). The programs include testing innovations in their six core countries in order to improve the effectiveness of their systems. Next, the organization uses research to understand how policies and practices can change to be able to advance services in developing cities. Subsequently, WSUP advisors, drawn from a pool of experienced professionals, are able to offer much-needed technical and infrastructure assistance. Finally, WSUP works together with city leadership in order to increase investment and resources for the necessary water services (WSUP, no date d). Naturally, in Madagascar, WSUP directed its investments to Antananarivo, with its large population and crowded urban center with little access to water and sanitation. The following section outlines how exactly WSUP and JIRAMA were able to spread water services to over 670,000 low-income residents in Antananarivo between the years 2010 and 2017.

Who is responsible for the increased water supply in Antananarivo?

To explain how the innovation process worked, it is important to first understand how the water system in Antananarivo is set up, and the structure of those who control all aspects of the system. JIRAMA, the state-owned utility, works closely with the Ministry of Water and Sanitation and is responsible for the supply of water and electricity to the entire island of Madagascar, currently providing services to over sixty operating centers which then deliver electricity and water supply to individual consumers, administrative departments, and industrial customers (Landitiana, 2014). The water is funneled from JIRAMA-owned plants into piping that reaches local communities throughout the city. Until recently, the lack of infrastructure that was able to bring water to these local communities had been the biggest obstacle facing JIRAMA, and there was an undeniable need for increased piping and water sources within the out-lying slum communities of the city. Naturally, these far-reaching responsibilities and projects make for extra challenges in the context of a developing country.

One of the overarching challenges plaguing JIRAMA and the Antananarivo government as a whole is corruption. Following the 2009 political crisis, state corruption remains a potent problem. Critics have blasted JIRAMA, particularly after periods of rain shortage which left water sources basically dry. In these instances, JIRAMA decided to ration water, supplying certain areas of the city with water for eight hours at a time (Keita, 2017). In addition to water rationing, electricity shortages from corrupt leaders diverting funds away from electricity supply leads to frequent power cuts for the citizens of Madagascar, resulting in large scale and sometimes violent protests against JIRAMA leadership and the Ministry of Water and Sanitation. Following a recent explosion of unrest, JIRAMA fired its director and replaced him. The city still experiences frequent power cuts, but leaders have acknowledged the issue and pledged to fix it. This is just one example of how the government in Madagascar and its capital city suffer from corruption. While it is difficult to ascertain the full extent of the corruption, eradicating the problem completely remains elusive (Keita, 2017).

Ultimately, despite the governance issues that JIRAMA and the Ministry of Water and Sanitation faced, the utility joined together with WSUP in order to improve water supply to 670,000 citizens of Antananarivo. Key partners of the initiative included Mayor Lalao Ravalomanana and staff of the Urban Commune of Antananarivo; the mayors and staff of the peri-urban communes of Greater Antananarivo; the Ministry of Water, Energy, and Hydrocarbon; and the Autonomous Maintenance Service for the City of Antananarivo.

How was this increased water supply implemented?

The framework for improving the water system investment was divided into two components. The first included working with target communities, particularly those urban communities that lacked adequate access to water, and more specifically, the water user associations, which are typically called communes. The second component included efforts to reduce the amount of water that is physically lost due to degraded pipes or intentional vandalism (Keita, 2017).

The first avenue which WSUP used to generate water coverage focused on local governments. The communes can be divided into smaller local government entities called *fokontany* (Keita, 2017). *Fokontany* are extremely important units in the government of Madagascar. *Fokontany* are the lowest subdivision of local government, and approximately 192 currently exist in Antananarivo (Landitiana, 2014). They are the decision-making entities for small communities, so they are deeply involved in the day-to-day activities of the more focused and confined community. These *fokontany* work with the water user associations that are responsible for water access activities and oversee what goes on within the water utility, most notably by collecting payments needed to pay JIRAMA on a monthly basis in addition to maintaining the water supply centers. The *fokontany* came together with water user associations and WSUP in order to determine where would be the most appropriate place within the community to have a water kiosk that is able to serve approximately five hundred people. Water kiosks have been used across the developing world, with particularly large-scale deployments in Zambia and Kenya, in order to provide safe, reliable, and low-cost water supply to low-income communities. In simpler terms, "a water kiosk is an outlet through which formal water providers deliver safe and reliable water at affordable prices to residents of low-income areas (Federal Ministry for Economic Cooperation and Development, 2009)." Water kiosks that are operated by formal water providers have been proven to be cost-effective solutions to combating unreliable and unsafe water supply that is found in many developing, low-income areas while also providing other essential community benefits including space to display important public health information and space to pursue commercial activities for vendors of health-related and agricultural products. A case study observing the water kiosks installed throughout Zambia even detected that around 50 percent of water kiosk operators were women, making this trade a quality promotion of income generation for women and thus increasing economic activity among the community as a whole. It seems likely that these economic and community benefits would reach kiosks throughout Madagascar, as well (Federal Ministry for Economic Cooperation and Development, 2009).

The physical picture of a water kiosk is a booth with usually around two or three water taps installed, usually set up in numbers among communities to ensure that residents do not have more than a five-minute walk to reach a kiosk (Federal Ministry for Economic Cooperation and Development, 2009). This allows the booth to be extremely accessible to maintenance, commercial activity, and general convenience. As has been discussed previously, the inclusion of the *fokontany* in deciding where the kiosks will be placed has been shown to improve the community's sense of ownership over the kiosk and prevent vandalism. The next step in implementing these new water systems involved WSUP and JIRAMA, in determining how much piping and construction would be necessary to reach those chosen kiosk locations. From there, the *fokontany*, the water user association, WSUP, and JIRAMA each provided a portion of the investment in order to begin construction on this kiosk and the necessary piping (Keita, 2017).

The second aspect of the WSUP framework involves preserving water from physical losses. Physical water loss is not just an issue in Madagascar, it is pervasive among even the most developed countries. For example, according to the U.S. Environmental Protection Agency, the average water loss in U.S. systems is 16 percent. These water losses are mostly due to pipe and infrastructure degradation; however, vandalism and theft also contribute to this issue, constituting a particularly potent problem in Madagascar. WSUP emphasizes this issue in all of its projects because it affects not only the amount of water provided, but also water pressure and quality of services. This duty also fell under the jurisdiction of the *fokontany*, which needed to be diligent about checking piping and infrastructure in order to decrease the amount of water loss throughout the city. With reduced losses, JIRAMA would be able to provide water to marginalized communities at lower cost, making it more likely for them to invest in new water kiosks, therefore more widely providing water services throughout the city (Keita, 2017).

Unquestionably, this framework to provide water supply to the 670,000 residents of Antananarivo draws largely on the city's underlying community structure. Key aspects of the capital city hub described in Chapter 1 of this volume did play a role. As Madagascar's gateway to the wider world, the city was able to garner significant resources from international donors like WSUP, and WSUP was able to harness JIRAMA, the Ministry of Water, Sanitation and Hygiene, and the *fokontany* to come together and invest the necessary resources to improve water and sanitation in the city to a more sustainable level. Having the largest population in the country in addition to government leadership residing within city limits certainly resulted in large donor organizations focusing on the city. Therefore, the city of Antananarivo was able to develop relationships with international networks in order to allocate resources to the country's largest urban area. However, as will be discussed further, there are key aspects of the capital hub that are missing in Antananarivo.

Is there potential to disseminate these water supply advances to the rest of Madagascar?

Is Antananarivo a suitable role model for future sustainable innovation in the rest of Madagascar? When looking at the basic facts, it seems questionable if Madagascar will be able to not only quickly, but effectively, provide clean water and sanitation to the other sixty municipalities on the island. With a shortage of strong governance, financial resources, and infrastructure, it appears that it would be extremely difficult for the rest of the island to innovate at such a high level as was achieved in the capital city.

Health concerns must be added to all the problems already discussed. In particular, Antananarivo and Madagascar as a whole have garnered a lot of international attention due to the prevalence of the bubonic and pneumonic plagues throughout the country. Between August 1, 2017, and November 22, 2017, the Ministry of Health of Madagascar reported a confirmed, probable, and suspected total of 2,348 cases of plague, which includes 202 deaths (an 8.6 percent fatality rate), according to the World Health Organization (WHO, 2017). Of the 2,348 total cases, 1,791 were pneumonic and 341 were bubonic, all others were unspecified (WHO, 2017). Only 22 percent of all cases were confirmed, 34 percent were probable, and 44 percent were suspected (WHO, 2017). The outbreak is somewhat contained, with detections in only 55 out of 114 districts (WHO, 2017). While this is certainly a setback for innovation in Madagascar, as resources need to be allocated to this top priority issue, this health crisis is not spread through waterways, so it should not inhibit future water supply improvement. Therefore, despite negative attention in the press, Madagascar's water system progress can still be seen as a role model for future advancement.

When looking at the work of Coca-Cola's Replenish Africa Initiative (RAIN), there is likely potential for similar techniques and frameworks seen in Antananarivo to be disseminated to other cities in Madagascar (Keita, 2017). One promising example comes from work that RAIN is doing in Toliara, in southwest Madagascar, and Mahajanga, in northwest Madagascar. RAIN uses a similar approach to WSUP in these areas, focusing on combining leadership from both JIRAMA and the local governing *fokontany* in order to provide better, cleaner, and more far-reaching water supply to a larger population, particularly in low-income areas (Keita, 2017).

There are two keys to being able to spread improved water systems across Madagascar. First, it is imperative that JIRAMA, the *fokontany*, and the donor network (i.e., WSUP, RAIN, USAID) are able to come together to demonstrate the model framework for increasing water supply throughout urban areas. The two-pronged approach discussed earlier used by WSUP in Antananarivo was carefully thought out and planned, and has been implemented in other cities in Kenya, Mozambique, Zambia, Ghana, and throughout Asia and Europe. This framework is also used by RAIN and other organizations because of its success in these other cities throughout the continent. Once the model framework is understood, the second key to spreading improved water systems can be undertaken; namely, that the municipality and all of the working actors mobilize investments in the project and try to replicate what was successful in Antananarivo. It is not easy to successfully accomplish these two steps, but based on the excitement, effort, and leadership seen in Antananarivo in recent years, it seems likely that resources will be disseminated across the country in order to continue working towards achieving the UN's Sustainable Development Goal 6, Clean Water and Sanitation. With the new government in place, there has been a call to focus on decreasing corruption, implementing innovations, and investing in infrastructure. These are all promising factors when looking towards the future of Madagascar as a country, and the continent's ability to improve and possibly lead the developing world in providing sustainable and innovative solutions to the issue of water management.

Conclusion

Madagascar, along with Sub-Saharan Africa as a whole, faces many challenges in achieving a future of sustainable, innovative progress in water management. Rapid urbanization, unstable political structures, and extreme effects from climate change all are among the obstacles faced by the entire continent. However, looking towards the future, there is promise that leadership will be taken in Africa in order to not only meet the UN's Sustainable Development Goals, but to provide the African people with clean, safe, and sanitized water systems.

While Antananarivo has seen considerable success in using its capital city status to garner international attention and aid, this case study finds that the city does not yet have a functioning capital city hub. What is missing is a network of professionals and scholars, a strong research community, profitable tourism, and a strong media mostly due to the lack of resources facing this developing country.

While bits and pieces of the hub exist, there are not yet enough of the components in place – and importantly, the reliable connections among them necessary to achieve a critical mass to support sustainable progress.

Despite this, Antananarivo has been able to capitalize on the use of the international community and national NGOs in order to innovate at a faster and stronger scale than the rest of the country, and these effective results are undeniably important to consider in forecasting the spread of sustainable innovation throughout the rest of the country.

References

Chaterjee, Rhitu. (2016). "Drought-Stricken Southern Madagascar Teeters on the Edge of Famine." National Public Radio, 5 December. www.npr.org/sections/thesalt/2016/12/05/504164887/drought-stricken-southern-madagascar-teeters-on-the-edge-of-famine.

Collier, Paul, Gordon Conway, Tony Venables. (2008). "Climate Change and Africa," *Oxford Review of Economic Policy* 24 (2): 337–353. https://doi.org/10.1093/oxrep/grn019.

Echart, Jochen, Kebreab Ghebremichael, Krishna Khatri, Harrison Mutikanga, Jotham Sempewo, Seneshaw Tsegaye, Kalanithy Vairavamoorthy. (2012). *The Future of Water in African Cities: Why Waste Water? Integrated Urban Water Management, Background Report*. Washington, DC: World Bank. https://openknowledge.worldbank.org/handle/10986/12275.

Editors of Encyclopedia Britannica. (2018). "Antananarivo: National Capital, Madagascar." *Encyclopedia Britannica*. 8 October 2018. www.britannica.com/place/Antananarivo.

Financial Times. (2016). "FT/IFC Transformational Business Awards Organised by FT Live." *Financial Times*. live.ft.com/Events/2016/FT-IFC-Transformational-Business-Awards.

Jacobsen, Michael, Michael Webster, Kalanithy Vairavamoorthy. (2013). *The Future of Water in African Cities: Why Waste Water? Directions in Development; Environment and Sustainable Development*. Washington, DC: World Bank. https://openknowledge.worldbank.org/handle/10986/11964.

Keita, M. (2017). "Interview with Malick Keita: Coca-Cola RAIN." 18 October.

Landitiana Soamarina Miakatra. (2014). "Community Participation and Water Supply in Deprived Areas of Madagascar." Field Actions Science Reports, Special Issue 11. http://factsreports.revues.org/3539.

Rondi, Luca, Sabrina Sorlini, Maria Cristina Collivignarelli. (2015). "Sustainability of Water Safety Plans Developed in Sub-Saharan Africa." *Sustainability* 7 (8), 11139–11159. 28 November 2017. http://proxygw.wrlc.org/login?url=https://search.proquest.com/docview/1729212662?accountid=11243.

Statista. (2017). "Madagascar – Urbanization 2006–2016 | Statistic." *Statista*. www.statista.com/statistics/455879/urbanization-in-madagascar/.

UN. (2015). "National Report Madagascar." *Habitat III*, May, pp. 1–43. habitat3.org/wp-content/uploads/Madagascar-English-National-Report-V1.1-Lite.pdf.

UNDP. (2009). "Madagascar." *Country Sector Assessments: UNDP GoAL WaSH Programme* 1, July, 6–20.

UNICEF. (2016). "Madagascar WASH Sector Key Results Framework (2016–2019)." *Madagascar WASH Sector*.

Water & Sanitation for the Urban Poor (WSUP). (no date a). "Antananarivo: Improving Services across Madagascar's Capital City." 8 October 2018. www.wsup.com/where-we-work/madagascar/antananarivo/.

Water & Sanitation for the Urban Poor (WSUP). (no date b). "Antananarivo Partnership Recognised for Water Access Transformation." 8 October 2018. www.wsup.com/blog/antananarivo-partnership-recognised-for-water-access-transformation/.

Water & Sanitation for the Urban Poor (WSUP). (no date c). "Approach: How We Work to Tackle Poor Water and Sanitation in Cities." 8 October 2018. www.wsup.com/approach/.

Water & Sanitation for the Urban Poor (WSUP). (no date d). "Madagascar: Driving Change in Low-Income Communities." 8 October 2018. www.wsup.com/where-we-work/madagascar/.

World Bank. (2011). "Vulnerability, Risk Reduction, and Adaptation to Climate Change: Madagascar." *Climate Risk and Adaptation Country Profile*. Washington, DC: World Bank Group. sdwebx.worldbank.org/climateportal/countryprofile/doc/GFDRRCountryProfiles/wb_gfdrr_climate_change_country_profile_for_MDG.pdf.

World Health Organization (WHO). (2017). "Plague – Madagascar." 27 November. www.who.int/csr/don/27-november-2017-plague-madagascar/en/.

9 Phnom Penh, an unsustainable capital city

Anthony Gad Bigio

Phnom Penh is one of the many Asian developing and emerging capital cities facing rapid economic growth and urbanization accompanied by overwhelming environmental and social negative externalities. This chapter attempts to demonstrate that in the absence of a political governance system which values the protection of the natural environment, public health and social inclusion, the multiple opportunities represented by the capital city hub are unfortunately neutralized.

When such values are upheld, urban development is more likely to be centered on infrastructure systems and basic urban services for the entire population: resilience and adaptation to climate change vulnerability, citywide drainage and sanitation systems, comprehensive solid-waste management, public transportation mobility, access to energy and renewable energy systems, secure land tenure and affordable housing opportunities, industrial pollution management, and protection and promotion of public spaces and of cultural heritage as shared assets. This "triple bottom line" of economic growth, environmental protection, and social inclusion, was the key approach of the Phnom Penh Green City Strategic Plan, prepared under the auspices of the Global Green Growth Institute in 2016.[1]

Cambodia's illiberal, quasi-totalitarian regime, in place for over 30 years, privileges instead the capture of wealth by the political elite and by associated national and foreign speculative investors. It is willing to sacrifice in return the balance of the urban ecosystem, the well-being of the resident population, and the overall sustainability of Phnom Penh (Strangio, 2014). In these conditions, the hub does not function as it does in more successful capital cities, because the national leadership arrogates all power to itself and prevents the other elements of the hub from participating effectively in the system through political repression. In such cases, the authoritarian nature of the regime overwhelms any benefits that the capital city can provide.

From master plans to privatized planning

Cambodia's economic boom and the government's embrace of market economics and foreign direct investment (FDI) since the late 1990s have profoundly marked the ongoing transformation of Phnom Penh and have projected it among

Figure 9.1 Evolving Phnom Penh skyline, including the landmark 39-story Vattanac Tower

the fast-developing cities of Southeast Asia. Such transformation has also been marked by huge private sector investments, destruction of natural resources and evictions of poor residents, in the context of little or no urban regulatory framework and of widespread corruption of public officials.

As a French colonial city, Phnom Penh was subject to European-style city planning, which started in the 1890s and continued until Cambodia's independence. While clearly designed for the interests and advantages of the colonial masters, the layout of the city – its infrastructure, public buildings, productive and residential developments – corresponded to an integrated vision, whereby land use was regulated and all investments would fit into an approved, predetermined plan. This planning practice continued successfully after independence in 1953, when under the influence of Vann Molyvann (the most prominent Cambodian architect-planner in charge of the Public Works department), Phnom Penh asserted itself as the urban image of a free and modernizing country, seeking deep connections to its past legacy of culture, religion and traditions as a source of contemporary inspiration. Phnom Penh during those times, fondly recalled as the Golden Era by many, earned its reputation as the "Pearl of Asia," and admiration of its urban development by regional leaders such as Lee Kwan Yew of Singapore, who visited in 1967.

The French tradition of high-quality public buildings (such as the Central Market, built in 1937) continued with the construction of the National Olympic Stadium, the Chatomuk Conference Hall, the Institute of Foreign Languages at the Royal University of Phnom Penh and many other buildings designed by Molyvann in the "New Khmer Architecture" style. The planning for the growing capital city continued for about 20 years after independence, following the pursuit of a balance between the public good, urban welfare, and opportunities for private and institutional investments in real estate. The first public housing complexes were also erected in Phnom Penh during that period.

The two decades of destabilization, genocide and destruction under the Khmer Rouge regime brought havoc to the capital city, which was forcefully evacuated. After the Vietnamese invasion and the civil war that ensued, the Paris Peace Accords of 1991 favored the gradual return of the resident population to the capital city. International aid to Cambodia included the reconstruction of the urban infrastructure of Phnom Penh. Establishing titles for both rural and urban properties was a major undertaking of the peace and reconstruction process of the 1990s, given the massive displacements of population and the deliberate destruction of title records that had occurred under the Khmer Rouge. In Phnom Penh, many properties had been informally occupied and their eventual re-titling favored current occupants, causing numerous conflicts and friction. Eventually, all privately owned property was re-titled, as well as private state properties such

Figure 9.2 Massive new buildings going up above the National Olympic Stadium

as public buildings. However, public state lands and natural resources such as rivers, lakes and wetlands remained in an unclear legal status, creating a favorable environment for encroachment and abuses.

At the government's request, the French Embassy in 2002 commissioned the preparation of Phnom Penh's Master Plan by the Paris metropolitan region's planning agency. The resulting 2007 draft Master Plan was prepared in close collaboration with the Phnom Penh Municipality (Royaume du Cambodge et al., 2007). However, its approval by the Municipality and by the competent Ministry was never ratified by the Council of Ministers. The lack of approval of the 2007 Master Plan has in fact introduced what has been defined as the "privatization of planning," whereby an all-encompassing, concerted and legalized plan for directing the future of the city was replaced by opportunities for negotiated concessions to large-scale national and foreign groups to invest in lucrative real estate deals and develop discreet and strategically located tracts of urban territory (Paling, 2012).

These gated communities, complete with their high-end on-site infrastructure, amenities, high-rise buildings and residential areas, offered the chance, in the government's view, to "modernize" Phnom Penh and place it among the successful Asian capitals. Such centrally located projects and edge-city developments include Diamond Island, Aeon Mall, Grand Phnom Penh International City, CamKo City, Boeung Kak Town and Chruy Changva City. The unregulated growth and redevelopment of the capital city has created significant pressure on urban systems, with the natural drainage and wetlands system as well as urban transport and mobility being perhaps the most affected.

"Privatized planning" has left Phnom Penh open to all sorts of lucrative redevelopment schemes with no provisions for the protection of its overall urban structure, zoning, heritage, natural resources or of its most vulnerable residents. It has been estimated that since 1999, about 150,000 people have been displaced from central locations within Phnom Penh to make room for such redevelopments (Strangio, 2014, 155). The evictions have occurred as the result of government-issued economic land concessions and of private real estate sales. As of 2014, 54 resettlement sites were located in the outskirts of the city, where displaced families often lack basic infrastructure and urban services, and are removed from employment opportunities (Strangio, 2014, 155).

In 2007, the government issued a lease over the Boeung Kak Lake, a key natural resource and part of the city's natural drainage system and protection against flooding. The lease was issued to a company belonging to the wife of a powerful ruling-party senator, acting as a front for a Chinese investment group (Strangio, 2014, 159). The 4,250 families residing on the lake borders were forcibly evicted (World Bank, 2011). The lake was back-filled with sand taken from the Mekong River banks and transformed into buildable land with an extension of 133 hectares and an estimated value of USD 1.3b. This was to be the location of the Boeung Kak Town development. It remains to this day one of the major urban environmental abuses of the regime, and the displaced families continue to manifest discontent and to demonstrate.

Economic land concessions, leading to 99-year land leases in favor of private developers, have paved the way for transactions involving "private state land" – i.e., public land under the control of public administrations. Many have proceeded with "land swaps" entailing the trading of valuable central properties in return for alternative premises in outer urban locations.[2] Modernizing Phnom Penh has consisted of the construction of new buildings, high-rises, malls, casinos, gated communities and high-end residences for the wealthy. In 2014, Aeon Mall opened with 100,000 m^2 of restaurants, boutiques and cinemas. Seven satellite city projects are scheduled for completion during the next 10–15 years, covering nearly 8,000 hectares, or 12% of the city's land area. During the first nine months of 2015, according to recent Ministry of Land Management statements, 532 new buildings went up in Phnom Penh, for a total of 4.5m m^2 and an estimated value of USD 1.7b.[3]

Urban skylines, such as the river view from the historic downtown area, have been permanently modified by the introduction of hotels and other buildings. Despite outcries from the United Nations Educational, Scientific and Cultural Organization (UNESCO) and civil society, Phnom Penh's urban heritage of colonial and modernist architecture from the 1950s and 1960s is being gradually demolished and replaced with non-descript contemporary buildings, reducing the city's attractiveness and memory.

In December 2015, the Council of Ministers approved a revised Master Plan, or "Strategic Orientation," which reflects the expanded perimeter of the capital city under municipal jurisdiction. It has not been made public or discussed with the interested stakeholders, and it remains in "draft" status.

In 2016, Phnom Penh seemed at a crossroads. The government could have made agreements with its international development partners favoring a more orderly and equitable development of the capital city, and acted on its Master Plan. Major aid investments were expected in urban transport, drainage and sewerage management, where donors were ready to support Cambodia with grants, technical assistance and large studies. The financing and implementation of those plans was only possible in the presence of a coherent and regulated urban development strategy, agreed to with other stakeholders.

The alternative scenario was the continuation of "privatized planning" and the unregulated flow of financing in infrastructure and construction, especially from Chinese sources, for lucrative deals. This is bound to negatively impact the long-term functionality of the city as a whole. A lack of improvement of its basic urban services and attention to the needs of low-income residents would bring Phnom Penh to a point of no return in terms of excessive vulnerability to flooding, traffic congestion, loss of urban productivity, water and air pollution, and inequality. These possibilities undermine the government's stated ambitions of Phnom Penh emerging as a rising Asian capital – but meanwhile fill the pockets of corrupt politicians and ruthless businessmen.

Elements of the hub and their interconnections

As noted in Chapter 1 of this volume, the capital city hub has little effect in situations of authoritarianism when the dominant political leadership blocks other stakeholders in society from effectively participating in the policy-making process. This is the case in Cambodia, which has survived under strongman rule for many decades. The specific roles and interplay of the key stakeholders in Phnom Penh's urban governance is described ahead.

Central and local governments

The Cambodian People's Party (CPP) has been in power since the UN-sponsored reorganization of Cambodia's political institutions following a peace agreement in 1991 which officially put an end to the decades of genocide, foreign occupation and civil war, and to the transitional single-party government. Despite the appearances of periodic democratic elections, the CPP has been in constant control of the country's institutions and its Prime Minister, Hun Sen, has acted as Cambodia's strongman since 1985. The liberalization of the national economy and its gradual opening up to foreign direct investment and globalization, which took place in the mid-90s, has not been accompanied by a further establishment of the rule of law or by a more open and inclusive society, but rather by the reinforcement of Hun Sen's regime which has widely profited from the economic growth.

The construction of a functional public administration and the progressive build-up of legislation and regulations have been largely the result of international support and pressure, as will be described further. In parallel, the CPP's and Hun Sen's personal control of the country's institutions has only increased, including over the army with its command of pristine forest areas and of highly profitable illegal logging activities. For 2017, Transparency International ranked Cambodia 161st out of a total of 176 countries surveyed, with a score of 21/100.[4] The recent political events seem to confirm that the country is heading towards a more totalitarian future.

The Phnom Penh City Hall, which had been instituted at the time of French Indochina, has the trappings of a democratically elected local government, but is in fact controlled by a set of governors appointed by the Ministry of Interior.[5] It is a weak and underfunded institution, as all infrastructure projects in the city are implemented by sector ministries, and its statutory prerogatives are quite limited. It is the national government that shapes the city of Phnom Penh, by making all the investment deals and retaining direct responsibility for the approval of construction projects over 3,000 m^2 (i.e., all large-scale developments), the location of industrial units and even the number of stories of high-rise buildings.

From a sustainability perspective, the government seems to follow a well-tested trajectory of "grow now, clean up later," as practiced by China and numerous other Asian countries in their ramp-up of industrialization and economic growth. Undoubtedly, the expansion of the national economy, based on export-oriented manufacturing and increasingly on speculative construction projects and tourism, has generated some redistributive benefits for the urban workforce, in the form of private sector jobs and of related greater access to the cash economy and to imported consumer goods.

Yet the urban battleground characterized by forced evictions of communities, highly unequal access to infrastructure and basic urban services, and the deteriorating urban environment, constitutes a potential weakness of the regime. The urban population of the capital city becomes more numerous, young people are more educated and access to information has been democratized and multiplied by the internet and by widespread mobile devices. Phnom Penh voters had already expressed their opposition to Hun Sen and his party in the 2013 political elections, and would have likely done so again in those of 2018 had they been given the opportunity. Instead, Hun Sen and his party won in elections that were neither free nor fair.

Civil society

The rather loose term of civil society encompasses, in the case of Phnom Penh, local community-based organizations and non-governmental organizations (some of which benefit from the support of international foundations and NGOs), independent media outlets, researchers and academics. Of particular relevance is Sahmakum Teang Tnaut (STT), a Cambodian urban NGO that provides pro-poor housing and technical assistance for housing and infrastructure in the low-income neighborhoods of the capital city (Doyle, 2012; Tudehope, 2012; Fukukzawa, 2014). STT also supports an information program, called "Urban Voice," through which communities report new events related to evictions, land takeovers and construction activities, which are occurring to the detriment of the urban poor. Its "Human rights based spatial planning" initiative, carried out as a joint project with People in Need, is funded by the European Union's Instrument for Democracy and Human Rights and by the Czech Development Agency. STT carries out mapping of urban displacement and high-quality analysis of urban transformation in Phnom Penh. People In Need (PIN) is a Czech NGO working in Cambodia on the promotion of tenure security for marginalized communities. PIN also supports disaster resilience in Phnom Penh, and produces high-quality technical assessments such as "City at Risk? Phnom Penh hazard, vulnerability and capacity assessment," published in 2015.

These NGOs in turn support community-based organizations, many of which regroup evicted residents from central city locations, from water ponds and canals that have been taken over by urban development. Waste pickers are supported by the Community Sanitation and Recycling Organization (CSARO), a Cambodian NGO committed to improving the environment and living conditions of urban poor in Phnom Penh. CSARO uses a participatory model to improve conditions in the peri-urban areas where waste collection is lacking. A special focus of its work is improving the living and working conditions of the thousands of adult and child waste pickers who make a subsistence living from Phnom Penh's waste. In 2006, CSARO purchased land for the construction of a new recycling and composting center in Roussey Keo district. Solid waste collection services were started in one new area in Dong Kao district and planning is underway for waste collection services in two new areas in Roussey Keo and Dong Kao districts.

In 2005, the Community Development Program established four new community development centers, and plans to expand activities to communities in five additional districts.

In terms of media outlets, Open Development Cambodia is a non-profit website which reports news and opinions about the initiatives of civil society, the government, and the international community actors present in Cambodia. The *Phnom Penh Post*, founded in 1992, and the *Cambodia Daily*, founded in 1993, are two independent English-language newspapers often critical of the government and outspoken in their stance. The *Cambodia Daily* was shut down in 2017 by the government on presumed tax liabilities. Both media outlets have reported consistently on the politics of urban renewal and development.

Private sector operators

Cambodia's economic growth over the past two decades has been built on private sector investments, mostly from regional and international sources (World Bank, 2015). Garment and footwear industries are key targets of FDI from China, South Korea, Japan and Vietnam. Investors are attracted by Cambodia's low labor costs, fiscal incentives provided by the government, and preferential trade agreements that the country enjoys with important export markets (USA, EU, Canada and Australia) (Asian Development Bank, 2014). According to the International Labor Office (ILO), there were 655 active garment and footwear factories in Phnom Penh as of June 2015. The garment industry is dominated by large firms (i.e., those employing over 100 workers). In contrast, the majority of Cambodia's small and medium enterprises are involved in the food and beverage industry.

The companies that build and own the factories are driven by bottom-line motivations and do not attach any concern to the health and well-being of the factory workers, or to the environmental impacts of industrial pollution. Labor and environmental legislations are in their infancy, and are perceived as threats to Cambodia's cheap production costs, and opposed by factory owners and their associations (Better Factories Cambodia et al., 2013).

On the other hand, a significant amount of the overseas orders are from EU- and U.S.-based well-established brands, which are under pressure and scrutiny from their own clienteles and stakeholders to improve their corporate and social responsibility. This pressure occasionally translates into requests from the buying corporations for the upgrading of labor and environmental conditions in the city's factories.

In the services sector, banking, financial and legal advisory services have been growing at a fast pace, with an overpowering presence of Chinese banks and capital, which dominate the banking sector. The major expansion of Chinese investments in the manufacturing sector, as well as in the infrastructure and construction sectors, is backed by highly concessional financing agreements that the government is obtaining from China. These are not conditional to any human rights, social or environmental considerations as those imposed by other important regional players in the region, such as Japan and South Korea, or by multilateral banks such as the Asian Development Bank and the World Bank.

Massive real estate investments are the result of FDI in the construction sector, with capital and expertise originating mostly from Japan, Malaysia, Thailand, Singapore, Hong Kong and China. While systematic data is scarce, the occupancy of the new upscale real estate is low, as it is beyond the reach of most Cambodians. Some of the developments might therefore be targeted to regional wealthy elites and the expatriate community, whose real numbers, however, do not justify the quantity of residential complexes currently being erected in Phnom Penh. Some projects are said to result from money laundering related to the profits from illegal logging occurring in the northern provinces, often at the hands of army generals. The ongoing real estate boom in the capital city clearly contributes to a strong urban economy and related job market, thanks to the downstream and upstream linkages of the construction industry. These are also present in the expanding stock of manufacturing plants, commercial parks and housing estates on the outskirts of Phnom Penh.

On the margins of this overall trend, there are some architecture firms that are expressing interest in green building design and in adaptive reuse of historic buildings. For instance, Re-Edge Architecture is promoting sustainable and green building design in its portfolio of buildings in Phnom Penh, which include the Galaxy Residence, Wangz Boutique Hotel and the Loop House. All structures have been designed with the purpose of reducing heat gain through passive measures such as smart layout, orientation of the building envelope and maximum use of natural ventilation. They make use of locally available materials to minimize the carbon footprint. Re-Edge Architecture is also interested in reviving the 1960s Khmer New Architecture by reusing existing historic buildings, which will reduce construction material costs and reduce construction waste going to landfills.

Eurocham Cambodia is an umbrella organization of European businesses operating in Cambodia and it promotes better commercial connections with European countries, especially with France, Germany and the United Kingdom. The United States Chamber of Commerce also has a branch office in Phnom Penh, and promotes better ties between U.S. businesses and Cambodia. Their importance is, however, becoming increasingly marginal, given how closely the country is gravitating into China's orbit.

International development agencies

Cambodia became a lower-middle-income country in 2015, according to World Bank classification.[6] Only Afghanistan, Nepal and North Korea are at this point the lower-income countries of Asia. Cambodia's low-income status, coupled with its troubled recent past marked by the Khmer Rouge regime and the related genocide, which caused the death of about two million people, were the principal causes of a significant amount of international assistance in the past few decades.

The United Nations provided significant assistance in the post-conflict period leading up to the Paris Peace accords of 1991 and in the few years that followed.

Reconstruction and development aid has been deployed primarily by the multilateral Asian Development Bank (ADB) and World Bank. The bilateral development agencies most present have been the Japanese International Development Agency (JICA), the Agence Française de Développement (AFD), the Gesellshaft für Internationale Zusammenarbeit (GIZ) and the Korean International Cooperation Agency (KOICA). JICA has a vast portfolio of urban projects focusing on urban planning, urban vulnerability, energy generation and distribution, transport infrastructure, public transit, built environment and manufacturing. ADB recently implemented a capacity development for urban management program leading to the preparation of a national urban development strategy. KOICA has been supporting technical assistance in urban planning, transport, energy and built environment.

A host of UN agencies have maintained a presence and work programs in Cambodia, such as UNIDO, which has been working on energy-efficiency initiatives in the manufacturing sector. UNESCO maintains active surveillance of the cultural heritage and historic built environment of Phnom Penh.

In more recent years, the Pilot Program for Climate Resilience (PPCR) and the Global Green Growth Institute (GGGI) have operated in Cambodia in their respective areas of work. A multitude of international NGOs have also fielded personnel in Cambodia in general and in Phnom Penh in particular, working in all areas such as education, health, agricultural extension, reforestation, demining, renewable energy, etc.

By statute, multilateral and bilateral development agencies support the government's work in policy formulation and in the implementation of investment programs. Financing and technical expertise are provided by the development partners, which also control disbursements for civil works and the procurement of equipment. Engaging international development agencies is well intentioned and aims at improving the provision of essential infrastructure, urban services and management capacity, and at assisting Cambodia with its economic growth strategy. However, the agencies have little say as to the institutional and political decision-making of the government, which may at times also undermine the effectiveness of their support. This puts the agencies in the morally ambiguous position of openly supporting a government whose purposes and practices contradict the goals of sustainable development they pursue. The 2007 privatization of Boeung Kak and the forcible expulsion of the 4,250 resident families caused the World Bank to interrupt its assistance to the Cambodian government for about ten years, in a rare case of withdrawal due to human rights violations.

The presence of international development personnel and of expatriates has been a significant characteristic of the life of Phnom Penh. This may be about to change, given the end of the concessional financing related to lower-income country status; the hardening of the regime towards its opposition, which may cause a more cautious approach from Western agencies; and the growing influence of Chinese regional presence and financing.

City sustainability assessment

This section provides an assessment of Phnom Penh's most pressing sustainability challenges.

Flooding vulnerability

Due to its location on the alluvial plain of the Mekong River, Phnom Penh is vulnerable to flooding, as it is surrounded by water at the confluence of the Mekong, Tonle Sap and Bassac rivers. This intersection of rivers is known as the "four faces" or locally as "Chaktomuk." Flood drainage is complicated by Phnom Penh's morphology, which is relatively flat: the elevation ranges from 4 meters in the Tumpum district to about 14 meters near Prey Key Village and the international airport. The water level in the rainy season can sometimes reach more than 10 meters, which can result in inundation in many parts of the city. Surveys have concluded that 30% of the capital area is lower than 8 meters, 45% lower than 9 meters and 60% lower than 10 meters (JICA, 2015a). Flooding can be caused by heavy prolonged seasonal monsoon rains, by recurring extreme storm levels, by seasonal high-water levels in the Tonle Sap and Mekong rivers or by a combination of these events. The destructive impacts of flooding are further exacerbated by uncontrolled property development, inadequate drainage and wastewater infrastructure, and the steadily growing unpredictability of the climate.

Cambodia has a history of life-threatening and damaging flooding events. Over the last 50 years, destructive flooding has occurred approximately every five years (People in Need Cambodia, 2015). Recent flooding and rainfall records suggest that the five-year flooding interval may be shortening. In 2011 and 2013, Phnom Penh experienced some of the most extreme flooding in history. The 2013 flood event was caused by a combination of heavier-than-normal monsoon rains, successive occurring typhoons, and rising water levels in the Mekong River. Over 17,000 families were impacted by the 2011 flood, and around 3,500 families by the 2013 flood in Phnom Penh (National Committee for Disaster Management and UNDP Cambodia, 2014). For the 2013 flood at a nationwide perspective, the total costs for damage and loss were about USD 356m, and a further USD 306m was required for recovery and reconstruction. The 2011 flood damage and loss costs amounted to USD 625m (National Committee for Disaster Management and UNDP Cambodia, 2014).

Flooding is becoming more serious in Phnom Penh. This is largely due to private-sector commercial developments in areas that were originally public green spaces, natural lakes or wetlands, and which historically and collectively served as natural flood control and wastewater management systems. During the period of 2003–2015, it is estimated that the area of Phnom Penh marshes, lakes and wetlands decreased by 50%. This observation is based on comparison of aerial photographs taken over this period (JICA, 2015a). The city has only about 70 hectares of green public space, mostly located in the inner-city core. The Phnom Penh inner core also has seven natural catchment ponds (Boeungs) to manage storm water.

Drainage systems constructed in the city core protect city properties to some extent, but outlying and peri-urban areas still remain at great risk to more frequently occurring flooding. Since 1998, Japanese International Cooperation Agency (JICA) grant assistance has helped to improve flooding in central parts of the city through a succession of projects. Earlier Asian Development Bank loan projects (2000–2003) also supported flood protection drainage in Phnom Penh. Based on its Master Plan on Drainage Improvement and Flood Control in Phnom Penh City, JICA has built a network of drainage channels, sluiceways, pumping stations and retention basins to control flooding in some densely populated areas of Phnom Penh. However, many densely populated parts of the city remain vulnerable to inundation after heavy rains, and are in urgent need of drainage infrastructure expansion and of rehabilitation of existing drainage systems (JICA, 2015b).

Sanitation, air pollution and public health

At many locations in the city's system of rivers and lakes, the water is extremely polluted, since Phnom Penh has no sewerage treatment system. Wastewater from homes and commercial enterprises flows directly into the storm drainage system, which eventually discharges into local rivers or marshes.

Water quality is monitored by the Ministry of Environment at 16 locations around the city. The parameters monitored included pH, total suspended solids, dissolved oxygen, biological oxygen demand, chemical oxygen demand, total nitrogen, total phosphorus and hexavalent chromium. A 2015 JICA study found that at a few of the monitoring points (Trabek pumping station and Tumpun pumping station), the dissolved oxygen concentrations were near zero, indicating that the waters could not support any type of healthy aquatic life. At these same locations, the concentrations of biological oxygen demand and total suspended solids exceeded the allowable limits by almost a factor of 30 in some cases. Total coliform content was also measured at the Tonle River, Mekong River, Bassac River, Prek Thnot River, Tamok Lake and Cheung Lake. Compared to water quality standards, the coliform content in these waters were up to a 100 times greater than the allowable limits (JICA, 2015a). In the low-lying peri-urban areas where wastewater is being discharged, there are reported high incidences of diarrheal diseases, hepatitis A, intestinal worm infections and protozoan infections resulting from fecal contamination linked with unsanitary living conditions and exposure to wastewater discharges (Otis, 2013).

In the main urban areas, most domestic and commercial wastewaters discharge to on-site septic tanks, which connect to the municipal sewer and drainage system, consisting of subterranean drains and of open canals. Surveys have concluded that about 70% of households have septic tanks that connect to the drainage system, while about 20% have septic tanks with no connection. Wastewater and surface drainage generated from central Phnom Penh drains to catchment lakes to the north or the south. Drainage flowing north is channeled to Boeng Pung Peay Lake before being discharged to the Tonle Sap River. Most of the city's wastewater (about 80%) flows south and is partially treated naturally via a network of wetlands, marshes and lakes. In both Boeung Tumpun and Boeung Choeung Ek lakes, nutrients from wastewater nourish the growth of morning glory fields, which are harvested by the local populations and sold in city markets. It is estimated that 20% of total vegetable consumption in Phnom Penh comes from harvesting gardens from these lakes. If not properly washed and cooked, vegetables grown under such circumstances could pose a public health problem (Otis, 2013).

Phnom Penh has no facilities for disposing of sludge accumulated from septic tanks. Previously, sludge originating from commercial buildings and households was extracted by sludge tanker trucks operated by private companies. These companies then disposed of the sludge in a lagoon at the solid waste dumpsite. Once the lagoon became overloaded with sludge, the dumpsite operators suspended the dumping of sludge. Since there is no current location to dispose of sludge, most sludge operators now illegally dump the sludge in drainage channels and in low-lying wetlands.

In the peri-urban and low-income areas, most households use pour flush latrines, which leach directly to local drains and watercourses. About one-third of households in extremely poor areas either bury sanitary wastes or practice open defecation, and there are reported high incidences of diarrheal diseases, hepatitis A, intestinal worm infections and protozoan infections resulting from fecal contamination linked to unsanitary living conditions and exposure to wastewater discharges (Flower and Fortnam, 2015). To adapt to frequent flooding events, the low-income households have resorted to building elevated housing platforms and walkways, or by moving to temporary shelters (Doyle, 2012).

Some citywide ambient air monitoring data taken from 2008–2012 at three locations for carbon monoxide, sulfur dioxide and nitrogen dioxide indicated that air pollution levels complied with regulations. A 2014 study ranked Phnom Penh very low in the air quality Environmental Performance Index (at 162nd out of 178 countries). This study concluded that air quality was unhealthy due to particulate matter or dust, and that the city had no ongoing real-time air monitoring capability for this parameter (Yale Center for Environmental Law and Policy, 2014). Another particulate matter monitoring program has observed that there is concern about the amount of dust in the city's air. The results of this program indicate that at some locations the concentration of respirable dust particles exceeds the World Health Organization standards by as much as a factor of 12. In the future, air quality in the city is expected to deteriorate further due to rapidly increasing vehicular traffic and the continued use of wood and charcoal for cooking purposes.

Solid waste management

Solid waste management in Phnom Penh is problematic, and waste collection in many parts of the city is haphazard – or lacking altogether. As a result, many tons of waste are routinely dumped into local rivers and ponds, are burned or remain uncollected. Uncollected waste often blocks local drainage channels and creates unsanitary conditions. Waste collection is unreliable or non-existent in the outlying or peri-urban areas where most low-income residents live. In a recent survey, it was reported that a majority of residents living in outlying Phnom Penh districts were not receiving any waste collection services. Where waste collection services are not available, residents reported that they either burned the waste or dumped waste on existing uncollected piles (Eang et al, 2015). Solid waste that is uncollected often accumulates in drainage channels and natural watercourses. During heavy rain events, the drainage capacity is limited due to the blockages from the waste. Forecasters predict that, due to climate change, more severe and frequent rainfalls will occur, and that solid waste collection will need to be greatly improved to ensure that drainage channels can function properly.

Reports indicate that domestic waste generation in Phnom Penh is about 600 tons per day. The waste is generated from households, commercial establishments and markets. Cintri Ltd, a privately owned company, has managed domestic waste collection and transport since 2002, on the basis of a contract with the Phnom Penh City Hall, which gave it exclusive rights for municipal waste management for 50 years. However, waste is not collected in many parts of the city since Cintri vehicles are unable to access certain parts of the inner-city core due to traffic jams and illegal parking near waste collection points, and it does not have facilities for servicing outlying peri-urban areas. In the past, Cintri drivers and trash collectors have gone on strike, leaving much of the city with accumulated piles of trash. Phnom Penh City Hall has also been criticized for the lack of private sector competition for waste collection services and for the lack of transparency regarding the Cintri contract.[7]

The waste dump at Steung Mean Chey, which had been in operation since 1965, has been nicknamed “smoky mountain” because of continuous trash fires as the transported waste is scattered or burned. The Phnom Penh Waste Management Division has described the existing dumpsite as a wall of trash measuring up to 10 meters in some sections. This facility has become overloaded with waste, and thus needs to be closed and decommissioned in an environmentally sound manner. In July 2009, a new landfill was opened at Dangkor Landfill with four waste disposal cells. The location is 15 kilometers from the city center and encompasses an area of about 26 hectares. It is reported that two cells are nearly filled to capacity. The landfill has quickly filled, and there have been observations that it is not operating in a technically sound and environmentally friendly manner. Proper waste compacting is not routinely practiced resulting in inefficient use of landfill capacity.

As many as five hundred waste scavengers, 80% of them women, make their living through informal recycling at the dumpsite. Children often accompany their mothers. It is estimated that waste scavengers earn around USD 5–10 per day. In October 2009, PPCH attempted to institutionalize the role of waste scavengers by defining their roles with respect to waste management, but this initiative did not result in any formal registration of waste pickers. Informal waste scavengers and recyclers also operate regularly at various locations within the city. Informal waste pickers can obtain recycling licenses through two recognized recycling organizations in the city.[8]

Urban transport

Phnom Penh has seen rapid growth in transport demand over the last decade, driven by rapid population growth and increasing economic activity. At the same time, improved incomes and cheaper imported vehicles have led to increased motorization and vehicle ownership. While the population of Phnom Penh grew by approximately 2.3% per year between 2001 and 2011, vehicle registrations grew by over 11.7% per year, more than tripling the number of registrations over the decade (JICA, 2014). Phnom Penh's transport is dominated by the use of two-wheeled vehicles. Motorbikes, para-transit (taxis, motorcycle taxis and tuk-tuks) and cars accounted for 52%, 14% and 10% of modal share in 2012, respectively. The use of public transportation is very limited, and there are no electrical or hybrid vehicles (Sieng, 2015).

The rapid expansion of the ownership and use of private transport in the city has resulted in increasing traffic volumes and increased congestion as infrastructure development and traffic management measures have been outpaced. The net result of this in the city has been slowing traffic and increased levels of congestion, increased road accidents and high levels of air pollution. Average driving speed in the city decreased from 22.9 km/h in 2000 to 14.6 km/h in 2012, indicating growing traffic congestion. At peak times, traffic speeds can be extremely low in central areas, at around an average of 10 km/h (JICA, 2014). Moreover, Cambodia now has the worst road traffic accident rate in Southeast Asia, with an estimated 2,000–2,100 fatalities in 2014. Motorcycle users accounted for 77% of the casualties and 68% of the fatalities, and the majority of motorcycle accidents occurred in Phnom Penh. The two leading causes of fatalities were speeding (51%) and alcohol abuse (18%).[9]

The provision of transport infrastructure has failed to keep pace with the rapid expansion of traffic volumes. Compared to other cities in the region, Phnom Penh has a low road density. Although in the central districts road density of 12.2 km/km^2 is similar to that of other city centers in the region, the suburban/peri-urban areas of the city currently have relatively low road densities. In regional cities where road densities are low (Hong Kong and Bangkok), this is largely offset by the provision of relatively good public transportation systems (JICA, 2014). Phnom Penh is currently without a system of public mass transit. Only three bus routes are currently in operation. Official figures of the extent of bus use are unavailable, but it appears to be low.

Ferry services are run by private companies and provide a link between Phnom Penh and the eastern suburbs across the river. Several wooden ferries operate from early morning to late evening; recently, a number of wooden ferries have been replaced by metal ones. There is limited information on taxis in operation in Phnom Penh. According to a recent study, there are two firms in operation, Tans-Choice Cambodia and Global Taxi (JICA, 2014). Trans-choice has about

80 taxis and employs around 200 drivers. It has plans to expand its taxi fleet to 300 during the next three years. This is likely to be an underestimation of taxis in operation, as many private-hire vehicles operate within the city.

Para-transit modes of transport are much more important in the city, particularly in central areas. Motodops, or motorcycle taxis, are unregulated, and there are no figures on their prevalence available. The operation of a private vehicle as a motorcycle taxi is often a secondary source of income, and there are likely as many in function as motorcycles in the city (i.e., 500,000). Although this is likely an overestimate, they are clearly an important means of transport in the city. However, the number of licensed three-wheel tuk-tuks is registered with the authorities, and these have increased dramatically over recent years. The same study estimates that there are approximately six thousand tuk-tuks in operation in the city (JICA, 2014). High levels of growth are expected to continue.

Industrial pollution

The manufacturing sector in the country and Phnom Penh is dominated by the textile, garment and footwear industries, which account for over 80% of manufacturing output, and in 2011 over 90% of employment in the sector. Nationwide these industries also employ around 80% of the non-agricultural labor force (607,000 people), 86% of whom are women. Garments and footwear are key export commodities, earning USD 4.9b in foreign exchange in 2013 (ILO and Better Factories Cambodia, 2013).

Pollution is a growing consequence of manufacturing located in the city. This includes pollutants to the air from energy production (including off-grid and backup diesel generation), solid and hazardous waste, noise pollution and possibly most acutely, water pollution. Although detailed figures on pollution attributable to the manufacturing sector are not available, ballpark estimates for water pollution were developed from water consumption figures available from Phnom Penh Water Supply Authority (PPWSA) for commerce and industry (JICA, 2015a).

These figures suggest that water pollution from these sources has grown rapidly, approaching 100,000 m^3/day by 2013, around 3.6 times wastewater emissions a decade earlier. Growth in water consumption – and by implication, wastewater production – has outstripped growth in either sector, suggesting that both sectors are making more intensive use of water. This may be due to structural shifts in the types of production and processes being undertaken, but nevertheless indicates the potential for a growing problem. It should be kept in mind that these figures also do not indicate the pollution load or toxicity of effluent emissions. While air and water emissions standards are in place in law, and there are requirements in law to treat waste products such that emissions standards are met, it is not clear that any meaningful monitoring or enforcement of pollution control measures takes place.

An additional issue compounding the problem with pollution discharges is the location of manufacturing industries in areas of the city with mixed uses, such as next to residential or agricultural land uses. In these circumstances, pollution emissions can cause significant problems for people living and working in nearby areas. This has been an important source of complaints to the city authorities. Similarly, the location of manufacturing industry in areas with unsuitable transportation access can cause congestion problems for the surrounding area.

With respect to the garment and textile industries, the key environmental problems are air pollution, GHG emissions from boilers, and potentially toxic wastewater from dyeing and bleaching processes. These industries are also very energy-intensive and typically operate using outdated and inefficient equipment. In addition, providing adequate work place ventilation and avoiding worker exposure to toxic chemicals are necessary to improving the working environment. Energy costs are said to be a significant burden for garment factories: even with discounted electricity rates, monthly bills reach hundreds of thousands of USD. The cost for charcoal and wood (used to fuel boilers to supply steam-irons and fabric dying basins) is also steadily rising. Selected factories have undergone energy efficiency audits, which found 30% energy savings are readily achievable via efficiency improvements to lighting, boilers, vents, sewing machines, compressors and other equipment (GGGI, 2014).

Urban energy use

Cambodia's energy supply mix is changing rapidly to meet the demands of urbanization, structural change and growth in the economy. Growth in the domestic power generation sector is a component. However, in the short to medium term, at least, Cambodia is likely to remain heavily dependent upon energy imports. Energy consumption is dominated by the residential sector, which accounted for around 68% of national total final energy consumption (TFEC) in 2013 (International Energy Agency, 2016). This is mainly composed of the use of traditional biomass for cooking and heating (91%), electricity (5%) and oil products (4%). The transport sector is the second largest consumer of energy overall (27%), and accounted for the largest proportion of commercial energy consumption, making up over 77% of imported oil products. Industry accounted for approximately 21% of national TFEC. Industry uses biomass to meet the majority of its energy needs (86%), the rest of which is composed of oil (9%) and electricity (6%). Among the energy-consuming industries, the garment sector is the leading consumer, followed by brick making, rice mills, rubber production and the food sector (particularly refrigeration) (Royal Government of Cambodia, 2013).

The residential sector in Phnom Penh differs markedly from the national picture. Figures from the 2014 national socioeconomic survey suggest that households in Phnom Penh rely to a much greater extent on modern commercial energy (liquefied petroleum gas [LPG] and electricity) relative to biomass for cooking than the rest of the country, and much more on grid-based electricity supply for lighting. However, these figures may overestimate the importance of modern energy sources. A more detailed survey conducted in 2013 found that charcoal continues to be an important energy source for cooking, and that households make more use of electricity for cooking.[10] The city accounts for the lion's share of electricity demand in the country, with peak demand of approximately 563 MW and electricity consumption of 3,268 GWh of grid electricity in 2014, which accounts for 67% the national total.[11] Phnom Penh receives grid electricity from a range of sources, including imported electricity from Vietnam, the hydro-electricity station at Kirirom, two thermal power stations and one diesel plant operated by EDC, and six other private producers (Tudehope, 2012).

As a consequence of high prices and unreliable supply, private backup diesel generators are a major source of electricity for high-income urban households, hotels and manufacturing firms (GGGI, 2014; Tudehope, 2012). An estimated 36% of firms in Cambodia use generators, with over 20,000 generators in Phnom Penh alone. This extensive use of diesel power has negative environmental and health impacts. Otherwise, access to electricity in Phnom Penh is relatively good, with 90% of households connected to the grid. However, according to Sahmakum Teang Tnaut (a local NGO), approximately 50% of Phnom Penh's poor settlements are not connected to the grid. Over 90% of Cambodians rely on combustion of traditional biomass for cooking and heating. Charcoal use is higher among urban households (25%, compared to 7% nationwide), and it is growing due to increasing urbanization and monetization of traditional fuels (GGGI, 2014 and Tudehope, 2012). The continuing reliance on combustion of traditional biomass for cooking and heating in urban and rural areas is resulting in high rates of acute respiratory diseases among Cambodians, particularly women and children. This remains an issue in Phnom Penh, despite the greater use of LPG and electricity. Moreover, the continued use of biomass in the residential sector and for industry is proving highly problematic for Cambodia's wider environment. Two fuel wood flow studies have shown that firewood and charcoal is sourced from distant surrounding areas including Kompong Speu and Kompong Thom, and that firewood extraction is leading to deforestation. In 2005, it was estimated that Phnom Penh was consuming 90,000 tons of charcoal extracted from the natural forest per year (GERES and Angkor Research, 2013).

Solar water heaters have a potential role to play in urban households; however, their application and use in Cambodia remains low (Matinga, 2012). Municipal solid waste and sewerage are other potential sources of bioenergy, but face the barrier of a dysfunctional waste disposal and collection system. Generation of biogas for heating purposes and electricity generation at water treatment plants and manufacturing plants has significant potential in Phnom Penh (Asian Development Bank, 2015). Despite high energy costs, there has been little movement towards more energy-efficient practices and technologies. The absence of energy efficiency standards and labeling means that consumers lack information on the efficiency of household appliances and other products. This puts more expensive energy-efficient appliances at a disadvantage, leading to proliferation of cheaper, inefficient products. Cambodia also does not currently have energy efficiency standards for buildings or vehicle emission standards.

Built environment

The built environment in Phnom Penh consists of a vast array of architectural styles and housing types. French colonial style buildings and villas historically used to dominate the cityscape. Now, many have been destroyed or are in a state of disrepair. Currently, typical housing units replicate the Chinese-style shop house, or consist of single-family homes. The shop house is the most predominant since it is less expensive and can be used for both commercial and room renting purposes. More recently, there have been significant investments in high-rise luxury apartments, entertainment complexes, commercial buildings and satellite cities. These types of structures will require large amounts of energy unless energy conservation and efficiency measures are integrated into construction and operation. Examples of new construction projects in Phnom Penh include Diamond City (Koh Pich), Bodaiju Residences and Booyoung Town.

A consequence of new construction projects is the forced eviction of residents originally residing legally or informally on the properties that undergo redevelopment. Many of the evicted families have had to relocate to impoverished peri-urban areas where proper housing is scarce. Evicted families in many cases have had to resort to building their own houses from recycled materials, bits of wood, scrap metal and old tarpaulin sheets. Such housing conditions have led to the formation of slum areas. Throughout Phnom Penh, there is an insufficient supply of affordable low-cost housing. Almost all new housing developments constructed by the private sector are targeted towards high-income earners and foreigners. The existing shop houses, while relatively low-cost, are not well designed with regard to energy efficiency and natural ventilation. Low-income laborers tend to rent rooms in shop houses, where overcrowding and heat stress can be hazardous to occupants. Furthermore, shop houses are also normally built right up to the sidewalk, leaving no space for green areas or trees.

A reported example of an existing green building design is the Phnom Penh Stock Exchange Building (Vattanac Tower, Figure 9.1) that was completed in 2014. The building is LEED certified and features low-energy insulating window glazing to control interior temperature, efficient lighting and cooling systems, and the use of natural lighting to reduce energy consumption. Other smaller scale commercial buildings have been built in Phnom Penh with energy efficiency and green building concepts. The Yellow Tower on Chroy Changva Peninsula features a living vertical wall garden that extends 50 meters in height. It is reported that the green façade reduces air conditioning costs by 30–50%.

The Phnom Penh city landscape will continue to be dominated by high-rise apartments (Figure 9.2), commercial complexes, gated communities and high-end satellite cities with limited or no occupancy. Availability of affordable housing will be insufficient in central Phnom Penh to meet the demands of many residents. More and more residents will continue to live in crowded inner-city shop houses or in slum areas located in peri-urban areas. With no guidance on constructing or retrofitting homes and building with regards to green design, energy efficiency and the use of locally available materials, electricity demand and GHG will continue to increase, with continued reliance on imported fossil fuels.

Public spaces and cultural heritage

Historically, Phnom Penh was a city rich in parks, green areas and cultural heritage buildings and sites. However, after the redevelopment of the city which began in the 1990s, many of the green areas and the city's cultural identity have been either destroyed to make way for new economic development or have been allowed to fall into disrepair due to inattention. The amount of green and open space in Phnom Penh has decreased markedly in recent years. Before the Khmer Rouge conflict, Phnom Penh was known for abundant green open areas that were naturally integrated with the local hydrological network of lakes and wetlands. Statistics now indicate that the total green and open areas in the city center amounts to about 0.36 km^2.[12] According to the JICA transport master plan, Phnom Penh has only 1.1 m^2 of green space per person, compared to New York City at 29.2 m^2 and Tokyo at 5.5 m^2 (JICA, 2014).

As for the state of public space in Phnom Penh, the Sisowath riverside promenade in the central district has been refurbished into an accessible, landscaped public space, which generates opportunities for urban recreation for residents and many visitors to the area. Various urban public parks, which had run into disrepair, have now been refurbished and present examples of some landscaping efforts. However, most public spaces are not well designed with regards to water management and natural cooling. Many inner district parks and gardens have been extensively paved with impermeable materials, and the trees and shrubs in these gardens have limited capacity for shading and for water conservation.

City sidewalks are entirely taken over by parked cars, motorcycles and informal vendors, making walking in Phnom Penh extremely difficult and dangerous. Many of the historic cultural heritage buildings and sites have been torn down to make way for new urban development projects or have been allowed to deteriorate. These building provide examples of architecture from the French Colonial era and also from the New Khmer Architecture movement of the 1950s and 1960s. Not only do these buildings provide cultural identity to the city, but they also have the potential for attracting tourists.[13] An example of a successful restoration project is the Phnom Penh Post Office, which was built in 1890 then restored and re-opened in 2004. The Post Office is a popular tourist attraction, while also being fully functional and offering services to the public.

State lands are likely to continue being sold off to private developers for short-term economic benefits. As a consequence, there will be fewer open public spaces for people to walk, exercise and to engage in socializing. As the city's population grows, there will be more competition to use the remaining limited public spaces. With fewer open spaces and greenery in the city, the impacts of climate change and increasing ambient temperatures could exacerbate the urban heat island effect and adversely affect the health of inner-city residents. The continued loss or disrepair of cultural heritage sites and buildings would result in a permanent loss of Phnom Penh's urban identity.

Conclusion

This chapter has tried to demonstrate that the opportunities offered by the capital city hub can exert a positive and innovative influence on the sustainability of a capital city only when the national political environment will allow it. At a time of rising authoritarianism in Asia and elsewhere, capital cities like Phnom Penh struggle to reach even minimum levels of sustainability. In the case of Cambodia, current geopolitics and the rise of China as a regional power in Southeast Asia counter the rise of an open society, in the absence of which the capital city will remain dominated by the ruling regime.

Recent political events confirm this scenario – 2017 was marked in Cambodia by the killing of government opposition members, the shutdown of independent newspapers, the arrest on trumped-up treason charges of the parliamentary minority leader and finally the judicial disbandment of the opposition party, whose parliamentary seats have been arbitrarily reassigned. The brutal repression took place ahead of the political elections of July 2018, which the ruling party dominated and outside observers declared neither free nor fair. In the absence of radical political reform, the capital city's environmental and social challenges can only worsen, notwithstanding the government's lip service to sustainability and green growth.

Notes

1 The Phnom Penh Green City Strategic Plan 2016–2025 was prepared by the International Center for Environmental Management (ICEM) for the Global Green Growth Institute (GGGI) as part of its technical assistance to the Government of Cambodia. I was the team leader for a group of international and national experts that carried out the assignment. I am indebted to my colleagues John Sawdon and Wayne Stone for their key contributions, and grateful to GGGI and ICEM for allowing me to draw from the Phnom Penh Strategic Plan report for the preparation of this chapter. The responsibility for the opinions expressed herein is, of course, my own.

2 Khy Sovuthy. 2014. Details Murky in Environment Ministry Phnom Penh Land Swap. *Cambodia Daily*, 28 February 2014; Chay Channyda. 2015. Land Swap Fears at Hospital. *Phnom Penh Post*, 17 February 2015.

3 Statement by H.E. Pen Sophal, Secretary of State, Ministry of Land Management, Urban Planning and Construction, on October 19, 2015 at the presentation of the National Urban Development Strategy Framework.

4 www.transparency.org/country/KHM.

5 http://phnompenh.gov.kh/en/.

6 According to the World Bank definition, middle-income economies are those with a per capita gross national income (GNI) of more than USD 1,045 but less than USD 12,736. Cambodia's per capita gross national income was USD 1,020 in 2014, USD 1,070 in 2015 and USD 1,140 in 2016 (latest data available).

7 Khy Sovuthy. 2015. Phnom Penh's Trash Problem Bigger Than Cintri, Company Says. *Cambodia Daily*, 28 February 2015.

8 Cambodian Education and Waste Management Organization. 2014. Institutional and Legal Framework of Solid Waste Management and the Development of the Current Landfill Operation and Management in Phnom Penh. Phnom Penh: Cambodian Education and Waste Management Organization.
9 Mom Kunthear. Road deaths to rise: minister. 2014. *Phnom Penh Post*. November 18, 2014.
10 A study by GERES and Angkor Research (2013) found that, for cooking fuel, Phnom Penh's residents use a combination of liquefied petroleum gas (LPG, main fuel for 38% of the population), charcoal (28%), electricity (27%) and firewood (7%).
11 EAC (Electricity Authority of Cambodia). 2015. Report on power sector of the Kingdom of Cambodia. Phnom Penh: Electricity Authority of Cambodia and EDC (Electricité du Cambodge). 2015. Annual Report 2015. Phnom Penh: Electricité du Cambodge. It should be noted that there are significant differences in the figures reported by the two agencies.
12 RGC. 2005. Urban Transport and Land Use in Phnom Penh. PowerPoint presentation, Bangkok, 26–27. September 2005.
13 UNESCO (United Nations Educational, Scientific and Cultural Organization) and Municipality of Phnom Penh. 2007. The Preservation of Urban Heritage in Cambodia. Report and Conclusions of the National Seminar held in Phnom Penh, 16–17 January 2006.

References

Asian Development Bank. (2014). *Cambodia: Diversifying Beyond Garments and Tourism. Country Diagnostic Study*. Mandaluyong City, Philippines: Asian Development Bank.

Asian Development Bank. (2015). *Renewable Energy Developments and Potential in the Greater Mekong Sub-Region*. Mandaluyong City, Philippines: Asian Development Bank. Better Factories Cambodia, International Labour Office, and International Finance Corporation. (2013). *Thirtieth Synthesis Report on Working Conditions in Cambodia's Garment Sector*. Geneva: ILO.

Doyle, Shelby. (2012). *Phnom Penh – City of Water*. Phnom Penh: Sahmakum Teang Tnaut.

Eang, Nhim Kim, Bour Chhayya, Pom Seila, (2015). *Solid Waste Collection and Management: the Forgotten Settlements in the Capital City of Phnom Penh*. Phnom Penh: Sahmakum Teang Tnaut.

Flower, Ben and Matt Fortnam. (2015). *Urbanizing Disaster Risk: Vulnerability of the Urban Poor in Cambodia to Flooding and Other Hazards*. Phnom Penh: People in Need Cambodia.

Fukuzawa, Meg. (2014). *The Phnom Penh Survey: A Study on Urban Poor Settlements in Phnom Penh*. Phnom Penh: Sahmakum Teang Tnaut.

GERES and Angkor Research. (2013). *Nationwide Domestic Use of Cooking Fuels and Devices: Baseline Survey*. Phnom Penh: GERES.

GGGI. (2014). *Cambodia Private Sector Analysis – Green Growth Opportunities Perspective*. Phnom Penh: Global Green Growth Institute. Unpublished manuscript.

International Energy Agency. (2016). *Cambodia: Balances for 2012*. Paris: International Energy Agency.ILO (International Labour Organization) and Better Factories Cambodia. (2013). Thirtieth synthesis report on working conditions in Cambodia's garment sector.

JICA (Japanese International Cooperation Agency). (2014). The project for comprehensive urban transport plan in Phnom Penh Capital City (PPUTMP). Final report – Executive summary.

JICA. (2015a). The study on drainage and sewerage improvement project in Phnom Penh metropolitan area. Progress Report II, November.

JICA. (2015b). Drainage improvement and flood protection in Phnom Penh – achievements, constraints and plans. PowerPoint presentation, July 2015.

Matinga, M.N. (2012). *Sustainable Energy for All – Rapid Assessment and Gap Analysis – Royal Kingdom of Cambodia.* Draft 1. Unpublished manuscript.

National Committee for Disaster Management and UNDP Cambodia. (2014). *Post-Flood Early Recovery Needs Assessment Report.* Phnom Penh: Royal Government of Cambodia.

Otis, D. (2013). Putrid lakes offer sweet relief to a city lacking water treatment plants. *Resilient Cities*, 18.

Paling, W. (2012). Planning a future for Phnom Penh: Mega projects, aid dependence and disjointed governance. *Urban Studies*, 49, 2889–2912.

Royal Government of Cambodia. (2013). *Draft National Policy, Strategy and Action Plan on Energy Efficiency in Cambodia.* Phnom Penh: Ministry of Industry, Mines, and Energy.

Royaume du Cambodge, Municipalite de Phnom Penh, Ambassade de France au Cambodge, Service de Cooperation et D'Action Culturelle. (2007). *Livre Blanc du développement et de l'aménagement de Phnom Penh. 2007.* Phnom Penh: Bureau des Affaires Urbaines, Municipalité de Phnom Penh, IAURIF.

Sieng, Trac Thai. (2015). Welcome to Phnom Penh smart city master plan. Presentation at the Sustainable Urban Transport Workshop, 8–10 July 2015, Danag, Viet Nam.

Strangio, S. (2014). *Hun's Sen's Cambodia.* New Haven: Yale University Press Silkworm Books.

Tudehope, Marcus. (2012). *A Tale of Two Cities: Review of the Development Paradigm of Phnom Penh.* Phnom Penh: Sahmakum Teang Tnaut.

UNESCO (United Nations Educational, Scientific and Cultural Organization) and Municipality of Phnom Penh. 2007. The Preservation of Urban Heritage in Cambodia. Report and Conclusions of the National Seminar held in Phnom Penh, 16–17 January 2006. Phnom Penh: UNESCO.

World Bank. (2011). *Cambodia Land Management and Administration Project, Implementation Completion and Results Report, ICR1491.* Washington, DC: World Bank Group.

World Bank. (2015). *Cambodia – Economic Update: Maintaining High Growth.* Washington, DC: World Bank.

Yale Center for Environmental Law and Policy. 2014. Environmental Performance Index. New Haven: Yale University. https://epi.envirocenter.yale.edu/ (On January 8, 2019, only the 2018 report was available on line.)

10 Conclusion

Missed opportunities

Robert W. Orttung

Capital cities have mixed results in terms of meeting their potential

Are national capital cities reaching their potential as hubs for innovation in pushing forward sustainability policies and practices? This book started with that question, and now it is possible to provide some preliminary answers. Judging from the quantitative and qualitative data gathered here, the answer is decidedly mixed. Some capital cities are clearly taking advantage of the resources at their disposal and pushing forward in a variety of ways with sustainability policies. Other capital cities are not living up to their potential for a variety of reasons. Overall, it is fair to say that capital cities need to do more to take advantage of their potential.

The concept of potential is popular in studies of education and labor markets, but do cities have potential? The answer is certainly yes. As Chapter 1 argued, capital cities bring together a hub of resources that differentiate them from other categories of cities. This powerful set of resources creates the potential for capital cities to lead the way in developing new kinds of sustainability policies. In this context, sustainability policies are those that promote economic development, concern for the environment, and social equity in ways that ensure that future generations will have access to the same or better life quality as current generations. Perhaps the best way to measure this potential is in its absence – frequently the citizens of a country express negative attitudes toward their own capital cities because they believe that the cities take too many resources from the country without providing results that the population appreciates.

What do the quantitative and qualitative data that we have assembled here tell us about how well capital cities are doing in achieving their potential? Chapter 3 provided a limited quantitative analysis of capital cities as a subset of the broader category of global cities by measuring their success in terms of minimizing carbon dioxide emissions. The research determined that in terms of achieving reductions in CO_2 emissions, capital cities that are also global cities do not distinguish themselves – they do not seem to have developed and implemented methods for reducing greenhouse gas emissions that go beyond what other cities are doing. However, OECD capitals did stand out in one way – OECD capital status implies an average reduction of over 2 tons of CO_2 per person per year. The most logical explanation for this finding is the active historical role of the OECD in promoting pedestrianization in urban areas. OECD capital cities rank high on the measure of walkability and use of public transportation, which is strongly correlated with lower CO_2 emissions. Viewed optimistically, this finding suggests that capital cities promote face-to-face contact among their residents, and this contact is important for the capital city hubs to work effectively. However, this interesting discovery does little to mitigate the overall finding that global capital cities do not perform much better than global cities as a whole.

According to a further series of calculations, the best performing sustainable cities in the world are San Francisco, Stockholm, and Seoul. While San Francisco is the major urban area in the Silicon Valley tech hub, Stockholm and Seoul are national capitals. The case study included in this volume explains Stockholm's success through the close links among the city's national government, local government, private companies, and research institutions. Overall, in terms of the limited quantitative analysis provided here, capital cities as a class do not stand out from the other cities studied, but they are among some of the leading cities in the world.

While the quantitative analysis provides a general sense of how well the cities meet their potential, the series of case studies in the second part of the book offers a more detailed analysis of how selected capital cities succeed – or fall short. Not surprisingly, Stockholm and Copenhagen are examples of the best performing cities. In Stockholm, the city government, the national government, research communities, and corporations are at the center of the discussion. They all work together effectively and have made Stockholm into a center of high-tech innovation. Copenhagen's public procurement policies make it a leader in using the power of the state purse to push private companies to promote sustainability goals. Local media and NGOs hold the key players accountable, and make sure that the suppliers are providing the kind of sustainable products that they promised.

Phnom Penh offers a stark contrast to Stockholm and Copenhagen. In particular, it demonstrates how authoritarian rule and the corrosive impact of corruption can prevent the hub from functioning successfully. Effective governance, based on democratic institutions and the kind of oversight that reduces corruption, are necessary to allow the various components of the hub to flourish on their own and interact in productive ways. Cambodia's illiberal, quasi-totalitarian regime instead merely created opportunities for the country's political elite and associated national and foreign speculative investors. The country's leaders were willing to sacrifice the construction and maintenance of a sustainable urban ecosystem in order to reap personal benefits from the system. In such cases, the authoritarian nature of the regime overwhelms any benefits that the capital city can provide. Notably, the 2007 privatization of the Boeung Kak area of the city and the forcible expulsion of 4,250 families living in that area caused the World Bank to take the rare step of interrupting its assistance to the Cambodian government for about ten years on the basis of human rights violations.

Naturally, most of the cities that we examined here fell between the extremes of the success stories and the real failure. These in-between cities show potential in a variety of conditions. In the context of Vietnam's single-party political system, whereby the government continues to hold considerable sway over public administration, expenditures, and safety, local NGOs were able to work with international organizations and the media to force the government to take action to protect local lakes and the tree canopy. Under pressure from numerous hub "spokes" – civil society organizations, universities, foreign embassies, international organizations, and private voluntary organizations, businesses, and media – the Hanoi city authorities instituted measures for public consultation, an innovation in what is otherwise an authoritarian political space. Effective use of the hub elements led the city government to become more responsive to public concerns and to increase the sustainability of the city.

One sure way for the capital cities to meet their potential is to develop stronger individual hub elements, better connections among these elements, and improved overall outputs from the entire system in terms of more sustainability policies. For example, when the national and city governments work in unison, the sustainability results are much more apparent. Stockholm clearly benefits from this convergence of values. In Washington, DC, on the other hand, where Republican national governments are often in conflict with the city's Democratic leaders, achieving sustainability progress is much more difficult, though some results are still possible.

The chapter on Cairo addresses an interesting question in this regard by asking whether hub functions can be improved by extracting them from an overcrowded mega-city where explosive informal growth means that many people do not have the necessary services, such as adequate housing and transportation that can move them quickly throughout the city, and transferring them to a new city that is purpose built to achieve sustainability outcomes. In this context, the choice confronting decision makers is whether to invest in "old" Cairo where all the elements of the hub already exist but do not have the ability to function effectively, or in creating a "new" Cairo designed to promote sustainability. Moreover, would the construction of the new city do anything to provide for the urgent social needs and problems of the current residents of the old Cairo, especially those living in the informal areas? As the discussion in the chapter shows, much depends on the how the new city is managed and what kind of relationship it will have with the old city.

The discussion of Madagascar's capital Antananarivo describes a city that has made some progress by using its capital city status to garner international attention and aid, yet ultimately concludes that the city does not yet have a functioning capital city hub. What is missing is a network of professionals and scholars, a strong research community, profitable tourism, and strong media, mostly due to the lack of resources that hinder progress in this developing country, as the chapter points out. While bits and pieces of the hub are in place, there is not yet enough of the components, and importantly, the reliable connections among them, necessary to achieve a critical mass to support sustainable progress.

Informal connections and collaboration matter more than institutions and wealth

In the most fundamental analysis, capital cities are different from other cities because they have more resources, so naturally we should expect more of them in terms of sustainability policies than we do of other cities. Chapter 2 provides extensive data to demonstrate that capital cities differ from other cities because they tend to have larger populations than they otherwise would have if they were not capitals. They also experienced higher rates of employment and income growth than other areas. Similarly, this analysis showed that they benefit from agglomeration economies and knowledge spillovers. Such data provide a strong foundation for our assertions that capital cities deserve more attention than they have received so far in the academic literature because they provide the potential for generating some of the most progressive policies in the sustainability field.

Additionally, the chapter showed that they exist in a variety of different formats – as federal districts, city-states, cities within a state, and cities within a unitary system. As Wolman notes, the literature suggests that city-state type of capitals in relatively rich countries are most likely to innovate. The city-state model has the most autonomy, and therefore potentially the most ability to innovate from an institutional standpoint. As the chapter pointed out: "With respect to the innovation generating capacity, it would seem, a priori, that, holding all else constant, independent districts would have less potential, particularly because they are most likely to have less autonomous local governments, while city-states would have the greatest capacity, since they are likely to have the most elaborated and dense set of institutions and network relationships." Some capital cities do not have the autonomy to make decisions on their own due to federal intervention, and this may limit the ability of the city governments to innovate sustainability policies. A central assumption in this analysis is that the capital city would have to introduce its innovations in the face of hostile forces that might not willingly accept the kind of innovations that it wanted to make. We will examine the autonomy and wealth hypotheses in turn.

The case study analyses provided in this volume suggest that it was the informal connections among the various elements of the capital city hub that proved more important than the institutional linkages. Also important, as noted earlier, is a convergence of interests among the various players in the hub, particularly the national and city governments. When you find that kind of intersection, which is visible in both Stockholm and Copenhagen (where it exists all the way up to the EU level), positive outcomes are more likely.

Adversarial relations existed to some extent in the example of Hanoi, but in that case, various elements of the hub were able to work together to nudge the city government in the direction of more sustainable policies for protecting the city's lakes and tree canopy. Cairo presents a more concerning case in this regard. There, some parts of the hub favored prioritizing resources for old Cairo while others suggested spending more on the new city. The private sector, planning community, and central government are the key stakeholders interested in building the new city, while local NGOs, international funding agencies, and academics, along with local government, have a strong interest in improving conditions in the informal developments of old Cairo, as the chapter points out. These differing interests among the various elements of the hub suggest that the policy outcomes might not be as mutually reinforcing as found in the case of the Scandinavian countries.

For the capital cities to be successful, is it simply a matter of rich countries producing more sustainable solutions while poor countries do not? By including case studies from northern Europe to southern Africa, we have wide variation in terms of the kinds of capital cities studied and the sustainability policy outcomes that they produce. The African country with significantly fewer resources, considerably weaker governance, drastically higher pressures, and greater environmental risks offers a useful relief to the more favorable conditions found in the Scandinavian cities.

Certainly, the cases of the Scandinavian and Cambodian capitals fit a narrative which argues that high wealth will lead to positive sustainability outcomes, while relatively poor countries are not able to produce at the same level. The rich Stockholm and Copenhagen were able to develop effective sustainability policies for developing high-tech industries, while the poor Phnom Penh became mired in authoritarianism and corruption that blocked any sustainability progress. However, the limited successes in the poor countries of Vietnam and Madagascar, where the capitals of Hanoi and Antananarivo were able to make progress in developing local water resources, shows that some cooperation and ingenuity can make up for a lack of resources. Similarly, the case of Washington, DC shows that even an abundance of resources in one of the world's richest countries does not lead to equally productive sustainability policies. While Washington has distinguished itself in terms of the construction of LEED buildings and bike sharing, it has not produced the kind of sustainability advances that one might expect, given its wealth.

Connectivity can be physical or virtual

Naturally, the capital city hub functions differently in each of the cities we examined. The first chapter posited a hub with eight distinct elements. However, none of the case studies were able to analyze all of these elements in full. The Stockholm case focused in on four – the national government, city government, private companies, and research institutions. The Copenhagen chapter also examined four elements – the city government and its connections to NGOs, the media, and research universities. Similarly, the Hanoi chapter covered five elements, stressing the city government in its relations with NGOs, the international community, the media, and researchers. The Antananarivo chapter stressed the absences of a strong media, research community, and tourism as crucial to explaining the limited capacity of that city.

By stressing the connections among the various hub elements, the cases emphasized the importance of the face-to-face connections in improving interactions among the various elements of the hub. But such personal, real world contacts were only the beginning of the process. In many cases, social media played a role in disseminating information and organizing people to take action even when they all lived in the same city. Social media was particularly important in Hanoi as civil society came together for environmental causes, but also in Stockholm, where it provided the basis of the growing tech industry. Investigative reporting by journalists and NGO reports also relied on social media distribution in order to reach audiences and have an impact. In this sense, capital cities are both actual physical places and centers for virtual connections on-line.

Not all elements of the hub are required to produce positive results. As we have seen, even in the most successful cases, only four or five elements of the hub were actively in play. Stockholm provides an example of a capital city in which the interactions of some of the spokes of the hub have been mutually reinforcing – and therefore resulted in a dynamic, entrepreneurial, and environmentally sustainable culture.

Strong hub elements produce the best results

One key finding of the case studies is that strong elements of the hub, working together, are necessary to create effective sustainability policies. In terms of the national government, the nature of the government matters for the eventual production of sustainability policies. In one of the successful cases, the Swedish government actively promotes sustainability policies and practices by investing in key infrastructure elements, such as highly efficient district heating for Swedish cities, hydroelectric power, levying a carbon tax, investing in new technology to reduce the overall risk involved, and ensuring the widespread early adoption of computer technologies, including among low-income members of the population. Its policies are typically coherent and coordinated across units. By contrast, in Hanoi, there was often confusion among the various government agencies as to who was responsible for the conditions of the lakes and ponds. The resulting mix-ups and neglect of responsibilities by competing agencies let developers encroach on the lakes and pollute them, diminishing the ascetic appeal of the city and cutting the environmental services that the lakes could provide. In this case, the coherent policies of the Swedish state opened the door more effectively for Stockholm's advances.

Even more important than the national government is the city government in each of the capitals. Of the case studies examined here, Copenhagen's use of public procurement policies to pursue sustainability goals was the most obvious example of effective governance at the city level. The chapter concluded that "the Danish capital successfully deploys innovative public procurement practices to integrate different parts of the hub as a way of implementing effective sustainability policies." While public procurement is not the first thing that most people think of when considering sustainability, it proved to be a powerful tool in advancing sustainability policies. The Copenhagen chapter shows how public procurement regulation can be used to incentivize diverse parties to support and advance sustainability goals and that it is a thread linking all the diverse constituencies of the hub described in Chapter 1 to develop sustainability solutions and create an environment where sustainability represents an important and relevant goal. Even though many of the regulations discussed in the chapter were developed at the EU level and were therefore exogenous to the capital city, it was the implementation by the city officials that brought those regulations to life and made them effective. Stockholm also worked to achieve a level of success by pushing ahead of the rest of the country; for example, by seeking to meet national sustainability goals ten years ahead of schedule.

Civil society plays a particularly important role in developing countries, where it can often compensate for state functions when the government does not have the necessary resources to perform functions effectively. In Cairo's Embaba informal development, a coalition of NGOs, academics, and journalists was able to improve water supply and quality levels. As the chapter noted, by engaging stakeholders in efforts addressing key shared concerns and with roles supported by new technology, specific weaknesses in the capital city could be addressed and improved. In that specific case, the city lacked data on water availability and quality and local residents were able to provide it using GIS and mapping applications. The role of civil society groups is equally important in developed countries, where they can hold governments accountable. In Denmark, DanWatch does numerous investigations of corporate sustainability practices in order to inform local governments and the public about the practices of the companies that supply the local government with products. Of course, the lack of resources hinders the work of NGOs. In Hanoi, for example, there are more than six hundred civil society organizations, but they are relatively weak, with small staffs whose members are often young and inexperienced. They depend heavily on international donors and thus are subject to those donors' priorities and constraints.

The local research communities and universities were crucial in providing support to the state actors in terms of delivering the know-how needed to produce effective sustainability policies. In the Danish example, the University of Copenhagen provided expertise regarding procurement practices to ensure that the suppliers were meeting their obligations. In Sweden, the policy of "professors' privilege" gave academics the rights to their inventions, thereby providing them with a strong incentive to commercialize their inventions on the local economy. The absence of a strong university was keenly felt in Antananarivo, which had to rely on outside expertise to compensate.

The media, both traditional and social, played a strong role in the effective cities. Media in Copenhagen conducted oversight on the government agencies. Through their work, they educated the public about procurement practices, and these reports, combined with the efforts of NGOs and universities, created public demand for sustainability policies. As the chapter pointed out, "Ultimately, the adoption of some sustainable procurement policies leads to a cascade of more policies as the first ones start to take effect. Similarly, exposing circumventions and discussing them in the media raises awareness about compliance issues and directs more public effort to bringing corporations back into line with their public procurement commitments." Similarly, in Hanoi, the media played an important role by helping to draw public attention to the situation with the lakes and ponds in the city and then disseminating the results of the lake research conducted by the Center for Environmental and Community Research, a local NGO working with international partners. In the case of the tree canopy, the media engaged actively in both raising the issue and facilitating direct actions to forestall tree removal. Social media played a role in sharing information and connecting the public, informing a broad section of users that state actors were removing trees and generating public interest in a planting movement to ensure that the canopy remained healthy.

As a source of wealth and knowledge, private corporations play a key role in the hub. The private sector is particularly prominent in Sweden, benefitting from strong support from the government and then giving back by building a collaborative entrepreneurial culture in Stockholm. The ICT Test Bed is one example where companies and the government work together to develop smart city technologies. As the Swedish chapter notes, within Stockholm's aspirational environment, ambitious individuals and firms have capitalized on opportunities to combine expertise in computer programming, engineering, and various technologies with a focus on environmental sustainability and a concern for the social implications of their business ventures in an effort to build commercially viable ventures, some of which have developed into unicorns, companies that are worth more than $1 billion. Despite its success, challenges remain in Sweden, where small startup companies often have difficulty marketing goods to government procurement officers and selling themselves to larger corporations.

The international community provides a variety of services to capital cities. Most prominently, international partners can be effective in helping colleagues in developing countries make progress. But, the international community is equally important for developed countries because embassies and other representations of international entities provide access for local actors to the international market. In Hanoi and Antananarivo, the international community provided strong support for environmental projects. As the Madagascar chapter points out, "Antananarivo has been able to capitalize on the use of the international community and national NGOs in order to innovate at a faster and stronger scale than the rest of the country." While all of the cities host some level of tourism, none of the case studies found a way to integrate this important part of the hub into our analysis.

Dissemination

Several of the chapters examined how well the capital cities would be able to disseminate their ideas to the rest of their countries. Stockholm found success in disseminating ideas, in part because the city and national government were closely interlinked in their thinking. The Vietnamese case shows partial success, with dissemination of ideas in the areas of lakes, but not in tree canopies. Madagascar is unlikely to be able to replicate the successful example of water delivery in the capital in outlying cities due to a lack of resources. Overall, the idea of dissemination is one that needs more research.

Alternative hypotheses

The introductory Chapter 1 raised several alternative hypotheses to the idea of a capital city hub. In contrast to the optimistic hypothesis that capital cities bring together a unique set of resources which make it possible to innovate sustainability polices, these hypotheses suggested that, in fact, capital cities do not innovate because they already have secure access to the national budget, the enemies of innovation are concentrated in the capital and can block progress, and that capital cities will lose out to global cities. What evidence can we now bring to bear in testing these hypotheses?

The case studies here show that even though it makes sense to speak of capital cities as a class of cities by themselves, they do not represent a monolithic group and display a wide range of effectiveness. Certainly, the case of Phnom Penh would meet the negative expectations suggested in these alternative hypotheses. There the leader used his control of the political system for personal gain. In this case, the enemies of innovation have certainly been able to win out.

In the more mixed cases of Hanoi, Antananarivo, and Cairo, there are certainly considerable battles going on between the advocates and enemies of innovation, but it is possible for advocates of change to build coalitions among the different hub elements to make progress in specific areas. Often the different hub elements bring to bear a winning set of resources that is needed to push sustainability efforts forward.

Finally, the cases of Stockholm and Copenhagen show that collaboration among various hub elements can lead to effective outcomes. In these cases, the different groups were able to work together for common purpose. Naturally, there were different interests involved. For example, the NGO DanWatch was able to expose some companies that were not living up to their sustainability commitments, and the media was able to red flag when state agencies did not follow through on their public commitments. These actors were able to overcome the opponents of innovation.

Future research agenda for capital cities

In making the argument that capital cities bring together a unique set of resources that allow them to produce innovative sustainability policies, this book has sought to develop interest in capital cities as well as their impact. In broaching this topic, it has certainly not answered all questions, and there are many opportunities for further research.

First, it would be useful to explore the opportunities that capital cities have missed and lay out in greater detail what they need to do to achieve their potential. This work would include identifying the areas where the cities have been most effective, and trying to replicate that experience in other locations and for other policy areas.

Second, there is considerable work to be done in terms of coming up with better measures for the performance of the cities. This book laid out a set of quantitative measures and a series of case studies. In the quantitative area, the measures so far have focused on the outputs of the cities, such as their level of greenhouse gases. It would also be productive to develop ways to measure their ability to create new policies. Also needed are more detailed measures of the components of the hub and ways to measure the amount of connectivity among the various hub elements. More effort is needed to understand the inner workings of capital cities and the nature of the networks among local, national, and international organizations/institutions, both public and private. Future work would benefit from a more sophisticated network analysis, including one that could take into account the various tensions in the network, including those within the capital city itself and friction between domestic and international actors. Such tensions were clear in the Phnom Penh chapter, for example, when the World Bank withdrew its support of Cambodia due to human rights violations.

Third, it would be helpful to gain a better understanding of how a city like Washington, DC fits into the broader picture of capital cities, since it is different from many other capitals in that it is not the economic center of the country. Other examples of such secondary capitals include Canberra, Brasilia, and Ottawa.

Finally, it would be useful to find new ways to determine how the alternative hypotheses discussed here could actually be put to the test empirically. How can we measure the real potential of capital cities and gain an understanding of how they measure up this potential?

There is a vast and rapidly growing literature on sustainability and policies aimed at increasing it. Hopefully, this focus on the role that capital cities can play in moving sustainability efforts forward will contribute to that literature and open new areas for investigation.

Index

For Product Safety Concerns and Information please contact our EU representative GPSR@taylorandfrancis.com Taylor & Francis Verlag GmbH, Kaufingerstraße 24, 80331 München, Germany